T0220406

Data Teams

A Unified Management Model for Successful Data-Focused Teams

Jesse Anderson

Apress®

Data Teams: A Unified Management Model for Successful Data-Focused Teams

Jesse Anderson
Reno, NV, USA

ISBN-13 (pbk): 978-1-4842-6227-6   ISBN-13 (electronic): 978-1-4842-6228-3
https://doi.org/10.1007/978-1-4842-6228-3

Managing Director, Apress Media LLC: Welmoed Spahr
Acquisitions Editor: Susan McDermott
Development Editor: Laura Berendson
Coordinating Editor: Rita Fernando

Cover designed by eStudioCalamar

Cover image designed by Freepik (www.freepik.com)

Distributed to the book trade worldwide by Springer Science+Business Media New York, 1 New York Plaza, New York, NY 10004. Phone 1-800-SPRINGER, fax (201) 348-4505, e-mail orders-ny@springer-sbm.com, or visit www.springeronline.com. Apress Media, LLC is a California LLC and the sole member (owner) is Springer Science + Business Media Finance Inc (SSBM Finance Inc). SSBM Finance Inc is a **Delaware** corporation.

For information on translations, please e-mail booktranslations@springernature.com; for reprint, paperback, or audio rights, please e-mail bookpermissions@springernature.com.

Apress titles may be purchased in bulk for academic, corporate, or promotional use. eBook versions and licenses are also available for most titles. For more information, reference our Print and eBook Bulk Sales web page at http://www.apress.com/bulk-sales.

Any source code or other supplementary material referenced by the author in this book is available to readers on GitHub via the book's product page, located at www.apress.com/9781484262276. For more detailed information, please visit http://www.apress.com/source-code.

Printed on acid-free paper

This book is dedicated to Sara, Ashley, and Grace.

Table of Contents

About the Author ....**xvii**

About the Technical Reviewer ....**xix**

Acknowledgments ....**xxi**

Introduction ....**xxiii**

Part I: Introducing Data Teams 1

Chapter 1: Data Teams 3

Big Data and Data Products 3

    The Terrible 3s, 4s, 5s… 4

    The Can't Definition 4

    Why Management Needs to Know the Definition of Big Data 5

Why Is Big Data So Complicated? 6

    Data Pipelines and Data Products 7

Common Misconceptions 8

    "It's All Just Data" 8

    "Isn't This Just Something Slightly Different from…?" 8

Why Are Data Teams Needed for Big Data? 10

Why Some Teams Fail and Some Succeed 11

The Three Teams 12

    Data Science 12

    Data Engineering 13

    Operations 14

Why Are Three Teams Needed? 16

    Three Teams for Small Organizations 17

What Happens If an Organization Doesn't Manage Correctly? 17

Chapter 2: The Good, the Bad, and the Ugly Data Teams 19

Successful Data Teams 19

What Big Data Success Looks Like 20

What Big Data Failure Looks Like 21

What Underperforming Projects Look Like 21

What Happens When a Team Is Missing 22

Figuring Out Value Generated 23

Issues with Scale in Data Science 24

Automate As Much As Possible 25

Part II: Building Your Data Team 27

Chapter 3: The Data Science Team 29

What Skills Are Needed? 30

Math 31

Programming 32

Distributed Systems 32

Communication 33

Domain Knowledge 33

Technical Debt in Data Science Teams 34

Hiring and Training Data Scientists 36

The Barriers to Retraining 36

Upgrading the Skills of Data Scientists 37

Finding Data Scientists 38

Meeting the Needs of Data Scientists 39

Introducing Software Engineering Practices 40

Too Much Process Stifles Progress 40

Chapter 4: The Data Engineering Team.. 43

What Skills Are Needed? 44

Distributed Systems 46

Programming 46

Analysis 47

Visual Communication 47

Verbal Communication 48

SQL 49

Schema 50

Domain Knowledge 51

Other Important Skills 51

Levels of Expertise 52

New Data Engineers 53

Qualified Data Engineer 53

Veteran 53

Further Specialization 54

Should the Data Engineering Team Focus Only on Big Data? 55

Common Misconception 56

Why Data Engineering Isn't Just Data Warehousing 56

Why a Data Engineer Isn't a Data Scientist 57

Why Data Engineering Is More Than Data Wrangling 58

The Relationship Between a Data Engineering Team and an Existing Data Science Team 58

Retraining Existing Staff 59

Software Engineers 60

SQL-Focused Positions 61

The Role of Architects 62

Placement in the Organization 63

Chapter 5: The Operations Team **65**

The Special Challenge of Operations on Distributed Systems 65

Job Titles for Operations Teams 67

What Skills Are Needed? 68

Hardware 68

Software/Operating Systems 69

Distributed Systems 69

Troubleshooting 70

Security 70

Data Structures and Formats 71

Scripting/Programming 71

Operationalization Best Practices... 72

Monitoring and Instrumenting... 72

Disaster Recovery... 73

Establishing Service-Level Agreements.. 73

Batch SLAs 74

Real-Time SLAs 74

Specific Service/Technology SLAs.. 75

Organization Code and Deployment SLAs 75

Typical Problems Faced 75

Your Organization's Code.. 76

Data and Data Quality... 76

Framework Software... 77

Hardware 77

Staffing the Operations Team.. 78

The Need for Specialized Training for Big Data 78

Retraining Existing Staff.. 79

Why a Data Engineer Isn't a Good Operations Engineer 80

Cloud vs. On-Premises Systems 81

Managed Cloud Services and Operations... 81

Chapter 6: Specialized Staff **83**

DataOps... 83

The Trade-Offs Made in DataOps... 84

Finding Staff for DataOps 86

The Value of DataOps.. 87

Relationship Between the DataOps and Data Engineering Teams...................... 88

When to Start Doing DataOps 90

Machine Learning Engineers 91

Finding Machine Learning Engineers 92

Where to Locate Machine Learning Engineers 93

Part III: Working Together and Managing Data Teams.....................95

Chapter 7: Working as a Data Team...............................97

Getting People to Work Together 97

  Staffing Ratios 98

  Should Data Teams Be Separate or Together? 98

  High-Bandwidth Connections 100

  An Iterative Process..100

Politics, Expectations, and Egos...101

  Data Project Velocity...101

  Create a Crawl, Walk, Run Plan 102

  Spreading the Love..103

  Communicating the Value of Data Engineering, Data Science, and Operations.................103

  Data Scientists Need the Contributions of Other Teams.................104

  Data Hoarding...105

Death by a Thousand Cuts 105

  The Curse of Complexity...105

  Going Off the Beaten Path 106

Technical Debt...107

Chapter 8: How the Business Interacts with Data Teams 109

How Change Can Be Accomplished 110

  Pushing from the Top Down...110

  Middle Pressure Up and Down 111

  Bottom Up...111

How Should the Business Interact with Data Teams?..........................112

  Case Study: Medical Insurance Domain Knowledge........................113

  Switching from Software as a Product to Data as a Product 116

  Symptoms of Insufficient or Ineffective Interaction 118

  Working with the QA Team 119

  Working with Project Managers 120

Funding and Resources 121

    Staffing 121

    Software and Hardware 121

    Cloud 123

Topics for Interaction 123

    Data Strategies 123

    Risks and Rewards 124

    Managing and Creating Realistic Goals 125

    Use Cases and Technology Choices 126

    Speaking Truth to Power 127

    Executive-Level Attention 127

Dealing with Uncertainty 128

    Data Sources Are Changing 128

    Output of Models Is a Level of Certainty 129

Don't Fear the Reaper's Data Products 129

Don't Forget the People Side 130

    Data Warehousing/DBA Staff 130

    SQL Developers/ETL Developers 131

    Operations 132

    Business Intelligence and Data Analyst Teams 132

Establishing Key Performance Indicators (KPIs) 132

    Data Science Team 133

    Data Engineering Team 133

    Operations 133

Chapter 9: Managing Big Data Projects 135

Planning the Creation and Use of Data Products 135

    One-Off and Ad Hoc Insights 136

    Exposing Data Products 137

Assigning Tasks to Teams 137

    Data Engineering, Data Science, or Operations? 138

    Collaboration 138

Rectifying Problems with Existing Teams 139

Effects of Using the Wrong Team for a Task 139

Long-Term Project Management 140

Technology Choices 141

A Mental Framework for Adding or Choosing Technologies 141

Reining in Technology Choices 142

Programming Language Support 144

Project Management 146

When It All Goes Wrong 147

Look Out for Nth Order Consequences 149

Chapter 10: Starting a Team 151

Establishing New Teams 151

The First Data Engineer 152

The First Data Scientist 153

Operations 153

The Critical First Hire 154

Location and Status of the Data Team 155

Embedding the Data Team in Business Unit 155

Hub and Spoke 156

Center of Excellence 156

Data Democratization 157

Getting the Team Right from the Start 158

Data Science Startups 159

Companies Lacking a Data Focus 159

Small Organizations and Startups 160

Medium-Sized Organizations 162

Large Organizations 163

What Should the Reporting Structure Look Like? 164

VP of Engineering 165

Working with Human Resources Departments 167

This Isn't Just a Focus on Titles 168

Adding New Titles 169

Focus on Technology Skills 170

Chapter 11: The Steps for Successful Big Data Projects 173

Presteps 173

Look at Data and Data Problems 174

Identify Business Needs 175

Is the Problem a Big Data Problem? 176

Decide on an Execution Strategy 176

Between Planning and Starting 177

Find Some Low-Hanging Fruit 177

Waiting for Data 178

Deeply Understand the Use Cases 179

Starting the Project 179

Getting Outside Help 179

Choose the Right Technologies 182

Write the Code 183

Create the Compute Cluster 183

Create Metrics for Success 184

Iterative Steps 185

Evaluate 185

When to Repeat 186

Chapter 12: Organizational Changes 187

Working with the Old Guard 187

What to Do with Previous Teams? 189

Innovation and Progression in Data Engineering and Data Science ... 190

Assessing Skills Gaps 191

Assessing Levels of Expertise 191

How to Perform a Gap Analysis 192

Interpreting Your Gap Analysis 193

Ability Gaps 194

Hardware Changes..195

    Cloud Usage...195

    Purchasing Clusters...195

    Systems for Big Data Teams...196

    Getting Data to Data Scientists..197

Chapter 13: Diagnosing and Fixing Problems.......................................201

    Stuck Teams..201

        The team says they'll get things done, but they said the same thing a month ago 202

        Whenever I give the team a task they come back and say it isn't possible......................202

        I can't tell the difference between the team making progress and being stuck................203

        We have trouble with small data systems, and we're having even more trouble with big data 204

    Underperforming Teams...204

        Whenever I give the data team a task, the team creates something that doesn't work 205

        Whenever I give the team a task, they create something that isn't really what I asked for....205

        The team can do elementary things, but they can't ever do something more complicated....206

    Skills and Ability Gaps..207

        How can I tell the difference between a skill gap and ability gap?.................................207

        Is the team having trouble learning how to program? 208

        The team is coming from a small data background and having difficulty learning big data and distributed systems 209

        My team says that you're wrong and that the task is easy..210

        Why can't I just hire some beginners and have them create this project? 210

        We built out a distributed systems cluster, and no one is using it 211

    Out-of-Whack Ratios...211

        My data scientists keep complaining that they have to do everything themselves...........212

        My data scientists keep dropping or stopping projects 213

        The data engineers' analysis isn't very good 214

    Project Failures and Silver Bullets 214

        We've tried several projects, and none of them went anywhere 214

        We've given up on big data because we can't get any ROI..215

        We've tried cloud and failed; now we're failing with big data 216

This is really hard; is there an easier way to get it done? 217

Is there any simple way or shortcuts to these problems? 218

We hired a consultancy to help us, but they aren't getting it done 219

We followed everything our vendor told us to do, and we're still not successful 219

We defined our data strategy, but nothing is being created 220

Our data science models keep failing in production 220

The Holy Grail 221

We copied someone's architecture, and we're not getting the same value 221

We have a really ambitious plan, and we're having trouble accomplishing it 223

The Software or Data Pipeline Keeps Failing in Production 223

We keep on having production outages 224

The data keeps on bringing our system down, and we can't stop it 224

It takes way too long to find and fix issues in code and production 226

Part IV: Case Studies and Interviews 227

Chapter 14: Interview with Eric Colson and Brad Klingenberg 229

About This Interview 229

Background 229

Starting Point 230

Growth and Hiring 231

The Primary Division into Data Science and Platform Teams 232

Bottom-Up Approach 234

Project Management 236

The Competitive Edge of Data 237

Advice to Other Companies 239

Lessons Learned in Hindsight 241

Chapter 15: Interview with Dmitriy Ryaboy 243

About This Interview 243

Background 243

Hiring at Twitter 243

Challenges of Data and Analytics 245

Task Ownership Structure 246

The Difficulty of Choosing Technologies and Creating Distributed Systems 247

Data Engineers, Data Scientists, and Operations Engineers 249

Project Management Framework 250

Business and Data Team Interactions 250

Keys to Success with Data Teams 250

Chapter 16: Interview with Bas Geerdink 253

About This Interview 253

Background 253

ING's Data Teams 254

ING Organizational Structure 255

Project Management Frameworks 258

Using Data Science in Banking.. 258

KPIs for Data Teams... 259

Advice for Others 259

Chapter 17: Interview with Harvinder Atwal 261

About This Interview 261

Background 261

Team Structure 262

Removing Barriers and Friction 263

Project Management Frameworks 265

Team KPIs and Objectives 265

Changes in Data Teams 266

Advice to Others 267

Chapter 18: Interview with a Large British Telecommunications Company 269

About This Interview 269

Background 269

Starting the Initiative 270

Working with the Business... 270

Data Scientists and Data Engineers 271

Supporting Models for the Enterprise 271

Moving from Proof of Concept to Production 272

Creating Enterprise Infrastructure and Operations 273

Project Management Frameworks 274

Advice for Others 274

Chapter 19: Interview with Mikio Braun 277

Dr. Mikio Braun 277

About This Interview 277

Background 277

Organizational Structure 278

Socializing the Value of Doing Machine Learning 279

Definitions of Job Titles 280

Reducing Friction 281

Project Management Frameworks 282

KPIs 283

Improving Data Scientists' Engineering Skills 283

Integrating Business and Data Teams 284

The Differences Between European and US Companies 284

Advice to Others 285

Index 287

About the Author

Jesse Anderson serves in three roles at Big Data Institute: data engineer, creative engineer, and managing director. He works on big data with companies ranging from startups to Fortune 100 companies. His work includes training on cutting-edge technologies such as Apache's Kafka, Hadoop, and Spark. He has taught over 30,000 people the skills needed to become data engineers.

Jesse is widely regarded as an expert in the field and for his novel teaching practices. He has published for O'Reilly and Pragmatic Programmers. He has been covered in prestigious publications such as *The Wall Street Journal*, CNN, BBC, NPR, Engadget, and *Wired*. He has spent the past 6+ years observing, mentoring, and working with data teams. He has condensed this knowledge of why teams succeed or fail into this book.

About the Technical Reviewer

 **Harvinder Atwal** is a data professional with an extensive career using analytics to enhance customer experience and improve business performance. He is excited not just by algorithms but also by the people, processes, and technology changes needed to deliver value from data.

He enjoys the exchange of ideas and has spoken at O'Reilly Strata Data Conference in London, Open Data Science Conference (ODSC) in London, and Data Leaders Summit in Barcelona.

Harvinder currently leads the Group Data function responsible for the entire data life cycle including data acquisition, data management, data governance, cloud and on-premises data platform management, data engineering, business intelligence, product analytics, and data science at Moneysupermarket Group. Previously, he led analytics teams at Dunnhumby, Lloyds Banking Group, and British Airways. His education includes an undergraduate degree from University College London and a master's degree in operational research from Birmingham University School of Engineering.

Acknowledgments

Thanks to everyone who believed in me. This book was a troubled birth. To my wife and kids. To my first editor Jared Richardson, may he rest in peace. A big thanks to Andy Oram.

Thank you to all the contributors: Ted Malaska, Paco Nathan, Lars Albertsson, Dean Wampler, and Ben Lorica. Thank you to the people who sat for case study interviews: Eric Colson, Brad Klingenberg, Dmitriy Ryaboy, Harvinder Atwal, Bas Geerdink, and an anonymous bloke in Britain.

Introduction

Welcome to my book. I'm hoping you're here to sort out your current team, or you're about to start a big data project. There's a long process ahead of you to fix or create a team—more like three teams.

If you're like me, you turn to the introduction to see what is going to be covered. This introduction serves as the brief synopsis of where my head is while writing this book.

This isn't the place to learn about the latest technologies or even old technologies. In fact, I'm going to deliberately eschew technologies and their discussion. They're always changing, and that's something you should know. You probably won't be getting 10- to 20-year life cycles out of these distributed systems technologies. They're often changing, and there's a new "best" technology right around the corner seeking to unseat the current champion.

With my consulting services, I've been able to work with many organizations and industries around the world. I was able to see the patterns and commonalities because I had access to a larger sample of data, and I could experiment on what the best practices should be. As you read the book, realize that every thought and idea isn't an academic theory. These were all firsthand experiences and truths that I learned, sometimes the hard way. I'll be sharing not just my stories and the story of my company, but the story of all the companies that I've worked with.

I'm primarily writing this book to establish and document the best practices for creating data teams. I've been on this quixotic adventure for many years. I got tired of organizations not using best practices and failing in their data projects. I started to research why these failures were happening. During that research, I found that few people had taken the time or effort to talk about their best practices. It wasn't management's fault that they were not following best practices because they were tribal knowledge! To fix this, I've been creating the body of work around technology and management for big data. I hope to help people prevent some of the failures.

When you're writing a book or article, sometimes the most insightful dialog comes during the feedback phase from the reviewers. The reviewers will read through a piece and give great feedback on a concept. This feedback can turn into a thread that is just as interesting as the piece, but isn't published. I've tried to bring reviewer comments into the book and subsections. That way, you can see what they said and their points of view. Some of these sections in the book will agree and expand on what I've said. Other sections will disagree with my ideas, and I respect the reviewers' points of view. There isn't just one path to success with data teams.

In keeping with my desire for many voices and opinions, I have case studies in the book. Instead of case studies of just companies, I did case studies on people. By studying people, I followed them through job changes and saw what they kept the same or changed as they moved to a new company. This allowed me to ask questions about how their unique experiences changed their strategy as they joined a new company.

For some, this book will be a confirmation of what they're experiencing, seeing, and trying to change in their own organization. They aren't being listened to, or their opinions aren't being heeded. This book will serve as the external validation of their views. When I speak at a conference or write a post, people often say that I write the things that they can't say or don't have the words to say. It's more like group therapy than a conference talk.

At the same time, other people are wondering why the rest of the world doesn't know the truths in this book because they seem glaringly obvious. The reason is that data team management is relatively new, and the differences aren't obvious. That's another reason I'm writing this book. I want to share what's evident to some of us and some of our opinions so you can make up your own mind.

I've been writing about data teams extensively for the past few years. Sometimes, I've written far more than I planned about a topic. Keep an eye out for these footnotes, and I'll give you the links to learn more about the topics if you're interested.

I didn't write this to be the definitive guide for creating data teams. This is the beginning of learning, and I'm trying to remove as many unknowns as possible. I expect you to use the information and know what comes next and some of the considerations as you make those choices. I want to share the questions you should be asking and the answers you should be looking for.

While I focus on distributed systems and big data, the same sorts of principles apply to small data. At small data sizes, the complexity will go down. This low complexity will move the technical bar for success to a more easily achievable level.

Finally, sit back, relax, and enjoy. Best of luck on your journey!

For more information and extras for the book, please go to www.datateams.io.

PART I

Introducing Data Teams

Before we go deeply into the details of each part of the data team and how to interact with them, I need to give you an overall introduction to data teams. Once you understand the basics of each team, we can start drilling down to the details.

CHAPTER 1

Data Teams

Oh I get by with a little help from my friends

—"With a Little Help from My Friends" by The Beatles

Making use of big data is a team sport. It takes several different kinds of people to get things done, and in all but the smallest organizations, they should be organized into multiple teams. When the help from these friends comes together, you can do some awesome things. When you're missing your friends, you fail and underperform.

Just who are these friends, what should they be doing, and how do they do it? This book answers these questions. The book covers many facets of forming data teams: what kinds of skills to look for in staff, how to hire or promote staff, how the teams should interact with each other as well as with the larger organization, and how to recognize and head off problems.

Big Data and Data Products

To make sure this book is right for you—that my topics correspond to what your organization is working on—I'll take some time to explain the kinds of projects covered in these pages.

How could we start a book about big data management without a definition of big data? What I want to do here is go beyond buzzwords and get to a definition that really helps management.

© Jesse Anderson 2020
J. Anderson, *Data Teams*, https://doi.org/10.1007/978-1-4842-6228-3_1

The Terrible 3s, 4s, 5s…

Everybody accepts that big data is a rather abstract concept: you can't just say you have big data because the sizes of your datasets hit certain metrics. You have to find qualitative differences between small and big data. That gets hard.

One of Gartner's original attempts to define big data led to the creation of the 3 Vs. Originally the Vs were variety, velocity, and volume. It's difficult for management to understand this definition. It was too broad. As a result, every company said their product was big data, and management still didn't understand the definition.

This led to people choosing their own definition. Pick a number between 3 and 20. That's the number of Vs that were defined.

Instead of providing clarity, these definitions really confused the issue. People were just looking through the dictionary for Vs that sound like they should fit. Managers were learning nothing that helped them manage modern data projects.

The Can't Definition

For management, I prefer the *can't* definition. When asked to do a task with data, the person or team says they can't do it, usually due to a technical limitation. For example, if you ask your analytics team for a report, and they say they can't do it, you probably have a big data problem.

It's imperative that the *can't* be due to a technical reason instead of the staff's skill. Listen to the reasons that the team says they can't do the task. These are some examples of technical reasons for a *can't*:

- The task is going to take too long to run.

- The task will bring down or slow down our production database.

- The task requires too many steps to complete.

- The data is scattered in too many places to run the task.

Obviously, your more technically trained people will offer a more precise and technical definition. I highly suggest you verify that your data teams really understand what is and isn't big data. If they don't, you could be relying on people who don't really understand the requirements.

Some organizations are smaller or are startups. What should they do, since they aren't saying "can't"—yet. The question should then be: will the organization have big data in the future? This future big data is really where many companies are focused. It's difficult to go back and reengineer lots of pipelines and code to use a different technology stack. Some organizations prefer to solve these problems from the very beginning instead of waiting.

Why Management Needs to Know the Definition of Big Data

It's crucial for management to understand what constitutes big data, and they should be guided by technically qualified people. This is because the mismatch of big and small data problems can really crush productivity and value creation.

Using small data technologies for big data problems leads to can'ts. Using big data technologies for small data problems is also a problem and not just because it leads to overengineering—it's creating major costs and problems.[1]

Just because many big data technologies are open source doesn't mean that they're cheap. Your costs will go up for infrastructure and salaries. Big data technologies tend to be pointy and filled with thorns, whereas small data technologies have fewer nuances that you'll have to fight. While small data lets you run all of your processes on a few computers, the number of computers explodes with big data. All of a sudden, your infrastructure costs are way higher. With big data, your response times, processing time, and end-to-end times could go up. Big data isn't necessarily faster; it's just faster and more efficient than using small data technologies for data that's too big.

Management needs to know when to push back on the hype for big data. Management shouldn't let engineers convince them to use something that isn't the right tool for the job. This could be resume polishing from the engineering side. But in the case of can'ts, big data could be the right approach.

Finally, if you're barely making it with small data technologies, big data technologies will be even more difficult. I've found that when an organization can barely exploit or productionize small data technologies, the significant jump in complexity leads to failure or underperforming projects.

---

[1] If you have small data, don't take this as belittling your use case, company, or project. Rather, you should be relieved that you've dodged the bullet known as big data (see www.jesse-anderson.com/2018/07/saying-you-have-small-data-isnt-belittling-your-use-case/).

Why Is Big Data So Complicated?

Big data is 10–15 times more complicated to use than small data.[2] This complexity extends from technical issues to management ones. Misunderstanding, underestimating, or ignoring this significant increase in complexity causes organizations to fail.

Technically, this complexity stems from the need for distributed systems. Instead of doing everything on a single computer, you have to write distributed code. The distributed systems themselves are often difficult to use and must be chosen carefully because each has specific trade-offs.

A *distributed system* is a task broken up and run on several computers at once. This could also mean data broken up and stored on multiple computers. Big data frameworks and technologies are examples of distributed systems. I'm using this term instead of talking about a specific big data framework or program. Honestly and unfortunately, these big data frameworks come and go.

Management becomes more complex, too, because the staff has to reach across the organization at a level and consistency you never had to before: different departments, groups, and business units. For example, analytics and business intelligence teams never had to have the sheer levels of interaction with IT or engineering. The IT organization never had to explain the data format to the operations team.

From both the technical and the management perspectives, teams didn't have to work together before with as high of a bandwidth connection. There may have been some level of coordination before, but not this high.

Other organizations face the complexity of data as a product instead of software or APIs as the product. They've never had to promote or evangelize the data available in the organization. With data pipelines, the data teams may not even know or control who has access to the data products.

Some teams are very siloed. With small data, they've been able to get by. There wasn't ever the need to reach out, coordinate, or cooperate. Trying to work with these maverick teams can be a challenge unto itself. This is really where management is more complicated.

---

[2]For a full explanation of this increase in complexity, I suggest you read `www.oreilly.com/learning/on-complexity-in-big-data/`.

Data Pipelines and Data Products

The teams we're talking about in this book deal with *data pipelines* and *data products*. Briefly put, a data pipeline is a way of making data available: bringing it into an organization, transferring it to another team, and so on—but usually transforming the data along the way to make it more useful. A data product takes in a dataset, organizes the data in a way that is consumable by others, and exposes in a form that's usable by others.

More specifically, a data pipeline is a process to take raw data and transform it in a way that is usable by the next recipient in the organization. To be successful, this data must be served up by technologies that are the right tools for the job and that are correct for the use cases. The data itself is available in formats that reflect the changing nature of data and of the enterprise demand for it.[3]

The output of these data pipelines are data products. These data products should become the lifeblood of the organization. If this doesn't happen, the organization isn't a data-driven organization and is not reaping the benefits of its investment in data.

To be robust, data products must be scalable and fault-tolerant, so that end users can reliably use them in production and critical scenarios. The data products should be well organized and cataloged to make them easy to find and work with. They should adhere to an agreed-upon structure that can evolve without massive rewrites of code, either in the pipelines creating the data products or among downstream consumers.

Data products usually are not one-off or ad hoc creations. They are automated and constantly being updated and used. Only in certain cases can you cut corners to create short-lived or ad hoc data products.

The veracity and quality of data products are vital. Otherwise, the teams using the data products will spend their time cleaning data instead of using the data products. Consistently low-quality data products will eventually erode all confidence in the data team's abilities.

---

[3]It's also worth noting other people's definition of data pipelines and related questions (see www.jesse-anderson.com/2018/08/what-is-a-data-pipeline/).

Common Misconceptions

Before we focus on how to make big data work for you, we need to dispel some myths that prevent managers from recognizing the special requirements of big data, hiring the right people to deal with big data, and managing the people effectively.

"It's All Just Data"

Sometimes people say it's all just data. They're trying to say there isn't a difference between small and big data. This sort of thinking is especially bad for management. It sends the message that what data teams do is easy and replaceable. The reality is that there is a big difference between small and big data. You need completely different engineering methods, algorithms, and technologies. A lack of appreciation for these differences is a cultural contributor to project failures.

There may be reasons that people think this. They may be in an organization with good data engineering. They're simply taking for granted the hard work of the data teams. From this person's point of view, it's all easy. This is one of the marks of a successful data engineering team.

Another reason may be that the organization doesn't have big data. They've been able to get by without having to deal with the complexity jump caused by big data.

"Isn't This Just Something Slightly Different from…?"

A common misconception with data teams is that they are only slightly different from a more traditional existing team in the organization. Managers think we're just going overboard with job titles and teams to make things more complicated. This sort of thinking contributes to failure because the wrong or unqualified team is taking the lead.

Business Intelligence

Some people believe that business intelligence is the same thing as data science. Yes, both teams make extensive usage of math and statistics. However, most business intelligence teams are not coding. If they are writing some code, they're not using a complex language. Put simply, a data scientist needs to know more than SQL to be really productive. They'll need to know a high-level and complex language to accomplish their goals.

Data Warehousing

Others think data engineering is the same thing as data warehousing. Yes, both teams make extensive usage of data. They'll both use SQL as a means of working with data. However, data engineering also requires intermediate-level to expert-level knowledge of programming and distributed systems. This extra knowledge really separates the two teams' underlying skills.

Although this book talks a bit about the continuing roles of database administrators (DBAs) and data warehouse teams, I'm not including their work as part of the data teams discussed in the book. The data products that they can create are just too limited for the types of data science and analysis covered in the book. Yes, some teams are able to create some products of value, but they aren't able to create a wide variety of data products that today's organizations need.

Operations

Another source of confusion is on the operational side. This also comes back to distributed systems. Keeping distributed systems running and functioning correctly is difficult. Instead of data and processing being located on a single computer, it is spread out over multiple computers. As a direct result, these frameworks and your own code will fail in complex ways.

The operational problems don't end with just software. Operations have to deal with data itself. Sometimes, the operational issues come from problems in the data: malformed data, data that isn't within a specific range, data that doesn't adhere to the types we're expecting—the list goes on. Your operations team will need to understand when a problem stems from data, the framework, or both.

Software Engineering

Finally, software engineering and data engineering look really similar from the outside. You might see a pattern here, but data and distributed systems make all the difference. Software engineering is no exception, and software engineers will need to specialize in order to become data engineers.

I've worked with software engineers extensively and throughout the world. Software engineering skills are close to data engineering skills but not close enough. For most of a software engineer's career, the database is their data structure and storage mechanism. Others may never have had to write code that is multiprocess, multithreaded, or

9

distributed in any way. The majority of software engineers haven't worked with the business in as deep or in as concerted way that is needed for data engineering.

The distributed systems that data engineers need to create are complex. They also change quite often. For most data pipelines, data engineers will have to use 10–30 technologies to create a solution, in contrast to the three or so technologies needed for small data. This really underscores the need for specialization.

Why Are Data Teams Needed for Big Data?

You might have checked out this book because you're part of a brand-new project, or one that's already underway, and have seen something amiss but can't put your finger on what's happening. I find this is often the case when someone is having problems with their data projects.

In this book, data teams are the creators and maintainers of data products. I have mentored, taught, and consulted with different data teams all over the world. After being brought on board, I have repeatedly encountered teams in various stages of failure and underperformance. The failure of data projects would prevent the business from benefiting or gaining any return on their big data investments. At some point, organizations would just stop investing in their big data projects, because they became a black hole that sucked up money while emitting little to nothing of value.

At risk of oversimplifying, I'll use the term *small data* for the small-scale work that organizations everywhere are doing with SQL, data warehouses, conventional business intelligence, and other traditional data projects. That term contrasts with *big data*, a buzzword we'll look at in the next chapter. The new big data projects seem familiar and yet also different and strange. There is an ethereal and hard to quantify the difference that you can't quite express or put your finger on.

This difference you're seeing is the missing piece that determines whether data projects are successful or not. Without this understanding that informs your creation of data teams and interaction with them, you can't be successful with big data. The key is getting the right balance of people and skills—and then getting them to collaborate effectively.

Why Some Teams Fail and Some Succeed

Years ago, I could only spot when a team was about to fail. I couldn't tell them what do to fix things, improve the team, or prevent failure. I felt quite powerless. I was seeing a train speeding toward a concrete wall, and I could only tell the passengers to get off before it crashed.

I knew that wasn't enough.

Figuring out the why, what, and who of these failures became somewhere between a challenge, obsession, and quixotic adventure for me. It became my research project. I absolutely had to figure out why very few teams succeeded and why so many teams that showed promise at the start ended up failing.

I became a success junkie, crazy-focused on people and organizations who were or said they were successful. When I found these people, they faced a barrage of specific and demanding questions: Why were you successful? How did you become successful? What did you do differently to become successful? Why do you think you were successful? Even as I'm writing this, I'm remembering people's faces as I asked these weird questions. The only way I could decipher this puzzle was by asking questions and assimilating other people's experiences.

I learned a lot from the projects that failed too. Encountering a project teetering on the edge of failure, I looked at the usual suspects but couldn't find anything obvious. Did people work hard? Yes. Were they smart? Yes. Was the technology to blame? No—at least not usually. I had to look deeper.

It was easy to blame the technologies, and that's what most organizations did after a failure. But I knew that was a total cop-out. Yes, there are problems and limitations with the technologies. Other organizations were able to productionize the same technologies, deal with the issues that the technologies had, and be successful. We're mostly talking about projects that failed well before the project ever made it into production or the first release.

No, these projects were failing so early in the cycle that there was a different culprit. Most of the time, the failures were the same over and over again. And yet, you couldn't really blame the staff or the management. There wasn't any body of work out to say why or what was happening.

Personally, I've been on this mission to create the body of work for management. I wanted to dispel the ignorance that led to all that waste. This book is a compendium of that effort. You'll see some of my related writings on the topic pop up as footnotes throughout the book.

The Three Teams

To do big data right, you need three different teams. Each team does something very specific in creating value from data. From a management's 30,000-foot view—and this is where management creates the problem—they all look like the same thing. They all transform data, they all program, and so on. But what can look from the outside like a 90 percent overlap is really only about 10 percent. This misunderstanding is what really kills teams and projects.

Each team has specialized knowledge that complements the other teams. Each team has both strengths and weaknesses inherent to the staff's experiences and skills. Without the other teams, things just go south.

We're going to go deeply into each one of these teams in Part 2, but I want to briefly introduce each one here. Like a dating show, let's bring on our three bachelors!

Data Science

Bachelor #1 likes counting things, math, and data. He's learned a little about programming. Meet the data science team!

When most managers think or hear of big data, it's in the context of data science. The reality is that data science is just one piece of the puzzle.

The data science team consumes data pipelines in order to create derivative data. In other words, they take data pipelines that were previously created and augment them in various ways.

Sometimes the augmentation consists of advanced analytics, notably the machine learning (ML) that is so hot nowadays. The members of a data science team usually have extensive backgrounds in mathematical disciplines like statistics. With enough of a background in statistics, math, and data, you can do some pretty interesting things. The result is usually a *model* that has been trained on your specific data and analyze it for your business use case. From models, you can get fantastically valuable information such as predictions or anomaly detection.

My one-sentence definition of a data scientist is:

> A data scientist is someone who has augmented their math and
> statistics background with programming to analyze data and
> create applied mathematical models.

At this initial juncture, there are few main things to know about data scientists:

- They have a math background, but not necessarily a math degree.

- They have an understanding of the importance and usage of data.

- They usually have a beginner-level understanding of big data tools.

- They usually have beginner-level programming skills.

This beginner-level skill is important to understand. We'll get deeper into this later—just know that this is a big reason we need the other teams.

A common mistake made by organizations just starting out with big data is to hire just data scientists. This is because they are the most visible element of big data and the face of the analytics created. This mistake is like trying to get a band together with just a lead singer. That might work if you're going to sing *a cappella*. If you're going to want any musical accompaniment, you'll need the rest of the band. A big focus of this book is to help you see and understand why each team is essential and how each team complements another.

Data Engineering

Bachelor #2 likes building model airplanes, programming, data, and distributed systems. Meet the data engineering team!

Going from data science in the lab to running data science at scale in the business isn't a trivial task. You need people who can create maintainable and sustainable data systems that can be used by nonspecialists. It takes a person with an engineering background to do this right.

The data engineering team creates the data pipeline that feeds data to the rest of the organization, including the data scientists. The data engineers need the skills to create data products that are

- Clean

- Valid

- Maintainable

- Usable at scale

- Sustainable for additions and improvements

Sometimes, the data engineering team rewrites the data scientists' code. This is because the data scientists are focused on their research and lack the time or expertise to use the programming language most effectively.

The data engineers are also responsible for choosing the data infrastructure to run on. This infrastructure can vary from project to project and usually consists of open source projects with weird names, or related tools provided by cloud vendors. This is a place where data teams get stuck, choose the wrong thing, or implement things wrong and get waylaid.

My one-sentence definition of a data engineer is:

> A data engineer is someone who has specialized their skills in creating software solutions around big data.

This team is crucial to your success. Make sure you have the right people with the right resources to guide you effectively. At this initial juncture, there are a few main things to know about data engineers:

- They come from a software engineering background.

- They have specialized in big data.

- Their programming skills are intermediate at a bare minimum, and ideally expert.

- They may be called upon to enforce some engineering discipline on the data scientists.

These data engineers are—at their heart—software engineers, with all the good and bad that comes along with it. They, too, need others to complement their shortcomings. So, in addition to interacting heavily with other teams, the data engineering team itself is multidisciplinary. Although it will be made up primarily of data engineers, there may be other job titles as well. These extra staff will fill out a role or skill that a data engineer doesn't have, or a task of lower difficulty that can be done in conjunction with a data engineer.

Operations

Bachelor #3 likes trains that run on time, hardware, software, and operating systems. Meet the operations team!

Running distributed system frameworks in production ranges from rock-solid to temperamental. Your code will likely have the same range of behavior. Who is responsible for keeping these technologies running and working? You need an operations team to keep the ship moving and everything chugging along.

Organizations accomplish their operational goals in two different ways.

The first is a more traditional operations route. There is a team that is responsible for keeping everything running. That team does not really touch the data engineer's code. They may have a hand in automation, but not in writing the code for pipelines.

The second is more of a practice than a team. This practice mixes data engineering and operational functions. The same team is responsible for both the data pipeline code and keeping it running. This method is chosen to prevent the quintessential "throw it over the fence" problems that have long existed between developers and operations staff where developers create code of questionable quality that the operations team is forced to deal with the problems. When the developers are responsible for maintaining their own code in production, they will need to make it rock-solid instead of leaving quality as someone else's problem.

Whether a separate team or a function of engineering, operations are responsible for keeping things running. This list of "things" is pretty and long, underscoring the necessity of operations. These things include

- Being responsible for the operation in production of the custom software written by your data engineers and data scientists (and maybe other people too)

- Keeping the network optimized, because you're dealing with large amounts of data and the vast majority of it is passed through the network

- Fixing any hardware issues, because hard drives and other physical hardware will break (less common in the cloud, but still occasionally requiring troubleshooting knowledge)

- Installing and fixing the peripheral software that may be needed by your custom code

- Installing and configuring the operating systems to optimize their performance

That list might sound like any operations, but let me add the things that really kick the big data operational team into overdrive. They must be:

- Responsible for the smooth running of the cluster software and other big data technologies you have operationalized

- More familiar with the code being run than usual, and understand its output logs

- Familiar with the expected amount, type, and format of the incoming data

My one-sentence definition of an operations engineer is:

> An operations engineer is someone with an operational or systems engineering background who has specialized their skills in big data operations, understands data, and has learned some programming.

At this initial juncture, there are few main things to know about operations engineers:

- They come from a systems engineering or operational background.

- They have specialized in big data.

- They have to understand the data that is being sent around or accessed by the various systems.

It's important to know there's really a different mindset between data engineers and operations engineers. I've seen it over and over. It really takes a different person to want to maintain and keep something running rather than creating or checking out the latest big data framework.

Why Are Three Teams Needed?

We've just seen the gist of all three teams. But you may still have questions about what each team does or how it differs. We're going to get deeper into the differences and how the teams support each other.

For some people or organizations, it's difficult to quantify the differences because there is some overlap. It's important to know that this overlap is complementary and not the source of turf wars. Each one of these teams plays a vital role in big data.

Sometimes managers think it's easier to find all three teams' skills shoved into one person. This is really difficult, and not just because there are so many specialized skills. I've found that each team represents a different personality and mindset. The mindset of science is different from engineering, which is different from operations. It's easier and less time-consuming to find several people and get them to work together.

Three Teams for Small Organizations

Small teams and small organizations represent a unique challenge. They don't have the money for a 20-person team. Instead, they have one to five people. What should these really small teams and organizations do?

What happens, of course, is that the organization asks certain staff to take on multiple functions. It's a difficult path. You'll have to find the people who are both competent to take on the functions and interested in doing so. These people are few and far between. Also, know that these people may not fulfill that role 100 percent. They can fill in for a function in a pinch, but they're not the long-term solution. You'll need to fill in those holes as the size of your organization grows.

What Happens If an Organization Doesn't Manage Correctly?

To round out this first chapter's overview, I want to share the effects of skipping some teams or placing the wrong people on them. Put simply, it condemns organizations to fail or get stuck at their big data projects. The specifics depend on complexity, use case, and what's missing. In any case, it leads the organization down a sad path.

I often see the second- or third-order effects of a missing team. I can see what is missing in the team by the explanation of the problem. I recognize what's missing because I've seen other organizations with the same problem. The root cause of the issue almost always falls into one of just a few categories. Sometimes, there are several layers to the problem, and you'll need to peel them back until you get to the roots of the problem. Only then can you really achieve the maximum potential from your data teams.

This book will help you get your organization and teams back on track or fix what's missing.

CHAPTER 2

The Good, the Bad, and the Ugly Data Teams

Successful Data Teams

And the dreams that you dream of
Dreams really do come true

—"Over the Rainbow" as sung by Israel Kamakawiwoʻole

Only 15 percent of businesses reported deploying their big data project to production, effectively unchanged from last year (14 percent).[1]

—Gartner Research

This quote from Gartner shows why, in the previous chapter, I focused so much on success or failure. The vast majority of big data projects either fail outright or underperform. The quote echoes my own experience in dealing with teams and projects. I'm really tired of seeing so many projects fail. At some point, organizations will just stop investing in their big data initiatives and groups because there is no payoff.

Obviously, your goal is to succeed, and you're reading to this book to either prevent or turn around a failure.

---

[1]Gartner Press Release, Stamford, CT, October 4, 2016, "Gartner Survey Reveals Investment in Big Data Is Up but Fewer Organizations Plan to Invest: Focus Has Shifted from Big Data Itself to Specific Business Problems It Can Solve: Analysts Discuss Analytics Leadership at Gartner Business Intelligence & Analytics Summit 2016, October 10-11 in Munich, Germany," http://tiny.bdi.io/gartnerfail.

© Jesse Anderson 2020
J. Anderson, *Data Teams*, https://doi.org/10.1007/978-1-4842-6228-3_2

As business people, we do case studies on businesses to see why they succeeded or failed. There are a plethora of books on this subject. However, with big data, we don't do that. There is a push to hide failures and never to talk about them. This lack of looking back at the failed projects just propagates the problem and causes the next failure. Until we learn from our mistakes, we're doomed to repeat them.

As you try to turn around a failing or underperforming organization, you will face the preconceived notion that nothing gets done and nothing will change for the better. It will be up to you to make the necessary changes. To help you see or show what could be done, I want to share some examples of getting value from your data teams and projects.

What Big Data Success Looks Like

Put simply, success with big data comes when value is created from your data. This means that the project is in production, solves a business need, and runs consistently. Each part of that previous sentence is actually a big achievement. Let's talk a little about each one.

Getting into production is a big deal. That means that the team has written enough code, created an architecture, and implemented things correctly. This really comes down to having the right, qualified team.

Solving a business need is also crucial. It means that the team worked with the business to identify a use case and actually provide value for it. This means that upper management stayed involved to make sure the data teams and business worked together correctly.

All of this work is moot if things don't run consistently. Success in this facet of the project means that each team wrote code that was of high enough quality to run in production. By extension, it means the code was unit tested and validated to work together. Finally, the operations team kept all of the moving parts running.

Companies that achieve success in these three facets of big data really pull ahead. To recognize success, look at a company that's leveraging its data and another that isn't. You should be seeing a quantifiable difference in revenue, customer satisfaction—both internal and external—decision-making, and strategy. A successful data project noticeably augments and propels the organization.

The actual value created will vary by industry and size of the company. It's hard to give a fixed amount or nonindustry-specific example. Also, this book isn't a business case study book.

What Big Data Failure Looks Like

Failure is easier to quantify and see than success. Projects never get into production or fail in production. Nothing is getting better. Deadlines come and go without actually creating the deliverable.

There are some consistent manifestations of failure. The most common is that the team is working strenuously hard. They're burning the midnight oil, but they're not getting anywhere. No matter how much time or effort they put in, the team doesn't seem to get something out the door. This just isn't fair to anyone.

Another manifestation of failure is being wholly unable to execute. The business comes to the team with an idea or requirement. The team responds that it is impossible or too hard to implement. The projects are dead before they even get started. As a direct result, only the easiest and simple projects get attention, and even those projects don't make it into production.

Some teams demonstrate failure by creating data products that have no value to the organization. To recognize this scenario, imagine the disappearance of your big data project. Would anyone care? Was anyone actually using the data, or was the cluster expanding to meet the needs of the expanding cluster?[2] If the big data initiative can be stopped and there would be little to no impact, that's a signal that the big data initiative is not creating value.

What Underperforming Projects Look Like

Underperforming organizations are more difficult to see—from the inside. To outsiders, it's clear when a team is stuck. The teams have just "hello world" or dead simple projects in production. Anything that is slightly more difficult never makes it into production.

Often, these teams are unaware they're underperforming. In my interactions with these teams, they're thinking they're really performing well or even outperforming other teams. After a closer look, it's clear that they're underperforming.

Underperforming and failing teams can be quantified in several different ways. How many data products are in production? How many of them fit enough of a business need to be used? How long is it taking to get something into production? How long ago did the

---

[2]Civilization fans will recognize this as an allusion to the Oscar Wilde quote, "The bureaucracy is expanding to meet the needs of the expanding bureaucracy."

big data project start, and how long did it take for the first data products to be released? Teams that take years are usually underperforming. Teams that never get something into production are failing.

Another manifestation of underperformance is a halt to adding new technologies. Although we don't add more technologies for the sake of resume building, they're frequently needed in healthy projects to improve value or handle new use cases. Teams need to add these new technologies because they improve or enhance their ability to accomplish the goal. If you're not adding new technologies, you're likely treading water. If you're starting a lot of new projects with old technologies, you're probably not making effective use of resources.

Another manifestation of underperformance is complaints from the business that they're not gaining value from the data. Successful organizations gain the maximum amount of value from their data, whereas underperforming organizations gain minimal to no value. You'll hear this complaint from the business because this criterion is what matters in the end.

Finally, take a look at how much you're spending on the data teams and hardware. How much ROI are you actually getting from your data, analytics, and insights? If you're not actually breaking even or even hitting opportunity costs, there's a major problem. More than likely, the team is underperforming relative to the amount of money spent.

What Happens When a Team Is Missing

To help people appreciate the contributions of each team under the umbrella of the data team, I ask people to imagine what data teams look like when that team is missing. For organizations that have already formed a data team lacking one or more of the teams I consider crucial, this isn't a hypothetical discussion; these are the problems they're experiencing on a daily basis.

- If the data engineering team is missing, there isn't anyone there to give an engineer's viewpoint about data product creation. In effect, the data scientists are doing all of the data engineering. The absence of a data engineering team doesn't mean that there's no data engineering being done. The burden simply passes to the data scientists, who aren't as capable of doing it. This leads to a mountain of technical debt that has to be fixed when the data product goes into production use, and that is costly to fix. The lack of the data engineering team inhibits the creation of data products that scale to the requirements of the business. It takes forever to get results from data products.

- If the data science team is missing, the organization's ability to create analytics is severely limited: analytics are just what a data engineer, business intelligence, or data analyst can muster. Only data scientists can do advanced analytics such as machine learning. Without the data science team, the organization probably can't create analytics complex enough to require programming.

- If the operations team is missing, no one can depend on the data products that are created. One day the infrastructure and data products are working, and the next day they're not. After waiting fruitlessly a long time for a dependable data product, the consumers will just stop using it because they can't rely on it. An organization can't base decisions on unreliable data products running on an unreliable infrastructure.

Only with all three teams in place can organizations really start creating value for the business.

Figuring Out Value Generated

It may be difficult to figure out the level of value generated by data products or data teams. When trying to measure that value, I don't ask the data teams themselves. I find that data teams have a rosier view of the value they generate than is warranted. For a better and more accurate viewpoint, I ask the business a few questions. I ask them to imagine that we are going to cancel the entire data project and fire the data teams. With these newly obliterated data products and teams, I ask them what their reactions would be. In response, I tend to get one of four general answers:

1. The first scenario is actually one where the data team is creating the most value. The business leaders will give a vehement, "No way!" The business is so opposed to making any changes to this lifeblood of data that's creating incredible business value. Making a slight change or removing the teams altogether would affect their day-to-day usage of data products and, ideally, decision-making. These projects and teams are creating extreme value for the business.

2. The second scenario shows a project that is creating minimal value. The business leader's reaction is "meh". Their ambivalence shows that the business isn't really using the data products on a day-to-day basis.

3. The third scenario shows a stagnated project that isn't creating any value. The business' reaction to a proposed cancellation is a snarky or pained "what project?". In such cases, managers promised the business that they could take advantage of data to make better decisions, but the data teams left this dream completely unrealized. The business has never had anything delivered into in their hands and couldn't ever achieve any value.

4. The fourth scenario is when a project is in the planning stages. The business has been promised new features, analytics, and the fixing of previous can'ts. There is a huge amount of anticipation from the business to finally get what they've been asking for. Now it's time for the data teams to deliver on these promises.

By looking at the business's reactions, I can get a clearer view of the business value being created without the reality field distortion of asking the data teams themselves. Every team should be aspiring to the first scenario's level of value generation. If your team is in the second or third levels, this is a warning sign that the team, skills, or another part needs work. At the fourth scenario, the team has to live up to its potential and deliver to the level of the first scenario.

Issues with Scale in Data Science

Part of what makes data science different from other analytics functions is the scale at which they work. This scale can mean different things for a data science team. It could mean that a data scientist needs to train their model or do discovery on 1 PB of data. It could mean that the model scoring needs to scale to work on hundreds of thousands of events per second. This scale requires that data scientists can't use the small data systems or technologies they may be used to. It sometimes means that their algorithms need to change from one that doesn't scale to an algorithm that does scale.

The sheer amount of data brings in issues that you don't hit at small scales. In 1 PB of data, it's possible—and more likely probable—that there will be bad data, data that doesn't adhere to limits, or is entirely malformed. This puts the data scientist's code in the spotlight because it will need to weed out the good from the incorrect data. You don't want a job that's run for 10 hours to stop because the data scientist forgot a check or didn't code defensively enough.

This scale is what brings distributed systems in the first place. If we stayed at a more modest scale, we could use the small data technologies and algorithms we've been using all along. This insight is based on the definition of big data in Chapter 1, which is crucial for management to understand.

Automate As Much As Possible

If your organization does moderate to extensive data wrangling manually, it may be a symptom that the data engineers lack programming skills or a missing tool in the data pipeline. If a task is done manually more than twice, you should really be looking at how to automate it. In general, look for every way possible to remove the humans from the loop. Automation will increase the reproducibility of results and infrastructure setup. Due to a lack of tools and programming skills, a data wrangler may not be able to fulfill this need.

If you take the proper steps within your company but still find that the teams are doing manual data wrangling or other steps, you may need to go back to the source of data and apply pressure to improve the original raw data or the exposing of the data product. This problem could come up when a data provider you partner with lacks a data engineering team to expose a proper data pipeline. Other times, poorly exposed data products could be an issue with an internal data engineering team that needs to be fixed. While some one-offs for manual steps are normal, too many one-off requests for the same manual steps are a sign that the data engineering team isn't creating the data products that your business wants or needs. This may expose a hole in your data product consumption that you need to fill. Instead of manual one-offs, a sufficiently technical team should be able to self-serve their own data from a properly exposed data product.

PART II

Building Your Data Team

The first management task in data analysis is to find people with the right skills, both technical and organizational. Teams require a variety of people with different skills—no one can be expected to do it all. This part of the book explains who you need to hire.

Many organizations hold on to the notion of a full-stack person who can do it all: machine learning, programming, and keeping everything running in production. These full-stack people do exist, but they're few and far between. You can spend a great deal of time looking for these people, and their salaries will be high.

There's also a big difference between doing it all and doing it all *well*. Some full-stack people can get by in a pinch, but by cutting corners on robustness, maintainability, or suitability for the business needs. It's hard for these people to see the blind spots in their own skills and abilities.

Sometimes, this desire for a single person comes from sheer budgetary needs: a small organization may decide they can afford only one highly trained person. The manager is forced to choose the one person they think will make the most impact relative to salary. The result is usually to hire a data scientist who is expected to accomplish everything.

Other times, the reluctance to diversify staff stems from the desire to avoid the difficulties of coordinating multiple people or teams. But relying on a single person is a path to problems and delays in the tough management parts of data teams that will have to be dealt with later. Creating data teams is hard work. The policy of avoiding a team doesn't scale. The solutions to creating data teams are held within these pages.

CHAPTER 3

The Data Science Team

And all this science I don't understand
It's just my job five days a week

—"Rocket Man" by Elton John

Of the three teams that make up a modern data team, we start with the data scientists, because they produce the output that their organizations use to make decisions. The other two teams exist to primarily work with the data scientists and secondarily with other parts of the organization.

The goal of the data science team is to create advanced analytics. On the high end of the spectrum, these could be generated by artificial intelligence (AI), where machine learning (ML) is currently the most common technique. On the low end, these analytics are built from advanced statistics and math. Other data team members may know statistics and math too, but not at the sophisticated level required for data science.

Thus, a data scientist combines an advanced math and statistics background with programming, domain knowledge, and communication skills to analyze data, create applied mathematical models, and present results in a form useful to the organization.

Some managers think that data scientists can create the end-to-end value chain of useful data, but that's not how big data pipelines work. Data comes in raw from many sources, such as web server logs, sensors, and sales receipts. The data engineers collect and arrange that data in a form that's useful for analytics, and the data scientists create the models and analytics that produce insights.

In this chapter, we'll talk about what a data scientist is, what skills they need, and get deeper into the differences between data scientists and data engineers. We'll also discuss how companies can recruit and train data scientists. Finally, we'll explain why data scientists must be able to deal with uncertainty.

Also, the data science team shouldn't be asked to create the data infrastructure or software architectures they use to create the data products, do discovery, train the models, or deploy these models. This is something more in the domain of the data

© Jesse Anderson 2020
J. Anderson, *Data Teams*, https://doi.org/10.1007/978-1-4842-6228-3_3

engineering team. Furthermore, data scientists are rare assets who have plenty to do besides creating infrastructure. Thus, the data science team should depend on the data infrastructure and software architecture created by the data engineering team.

This chapter is based on Paco Nathan's work on building data science teams.[1]

What Skills Are Needed?

The skills needed on a data science team are

- Math

- Programming

- Distributed systems

- Communication

- Domain knowledge

These are very general terms. We'll look in the following subsections at what they mean specifically for a data scientist, and why they're important.

We'll see in the following two chapters that the data engineering team and operations teams are multidisciplinary. There may be several job descriptions. In contrast, the data science team consists only of people with the job title of data scientist—although they can divide up skills and have different levels of skill. Some organizations expect to place their data analysts and business intelligence users on the data science team, but that's not an effective combination. The data analysts and business intelligence users are consumers of the data products created by the data science and data engineer teams.

DATA ANALYSTS AND DATA SCIENCE

Not every problem requires the demanding level of technical and machine learning that a data scientist provides. In these situations, a data scientist's time can be overkill. However, there's still a need for descriptive and diagnostic analytics, and data analysts can use big data

---

[1]You can view Building Data Science Teams for his take: https://learning.oreilly.com/videos/building-data-science/9781491940983.

technologies to answer these questions. By using data analysts on these problems, we're able to remove the constraints of having few data scientists and use the more available data analysts.

Where do data analysts fit into this new data teams organizational model? At Moneysupermarket we actually combine data scientists and data analysts into cross-functional teams. The teams can deliver much more without help from other teams for their domain. It works very well to speed up delivery, spread knowledge, and cross-train. This cross-pollination from the data scientists to the data analysts improves the data analyst's technical and analytical skills. We've seen the move from data analyst to data scientist as a promotion path as data analysts improve their programming and machine learning skills.

—Harvinder Atwal, Author, Technical Reviewer, and Chief Data Officer, Moneysupermarket

Math

At the very least, a data scientist should have completed Algebra 2 or the uppermost high school mathematics qualification in other parts of the world. This level of math would provide the very base level of math skills. At the high end, a data scientist would have a Ph.D. in statistics, mathematics, or a field with extensive math backgrounds. The majority of data scientists I've interacted with have PhDs. That isn't to say non-PhDs can't be a data scientist—but they'll really need to improve their math skills.

Because university programs that focus on data science are relatively young, many data scientists come from other technical backgrounds that are heavy in math and statistics. Some disciplines that I've seen produce successful data scientists include economists, physics, astronomy, rocket science, and biological or physical sciences such as neuroscience or chemistry.

If an entire data science team has only completed Algebra 2 as their highest level of math, the data science team will be limited. The team will have a basic understanding of the math requirements, but won't be able to progress to the next level of machine learning that requires an in-depth mathematics background.

Programming

Most data scientists have beginner-level to intermediate-level programming skills. Programming prowess and languages are highly varied. In general, data scientists program in Python and, to a lesser extent, Scala. Finally, some data scientists program in R (this is more difficult to use at scale in production).

I've found that the majority of data scientists are self-taught in programming. If they've taken a class on programming, it's been an introductory class. It is quite rare for a data scientist to have a hard-core computer science or software engineering background. Rather, they usually picked up programming as a means to an end, an ad hoc way to complete a task that the budding data scientist couldn't do otherwise.

A common reason for informally learning programming is that, during their Ph.D. thesis, they need to program something. This required a push into programming that they enjoyed, and they started going even deeper. Before you know it, they're marketing themselves as a data scientist.

Continuing with this theme, some people had a task that required processing a massive dataset. Once again, this learning was a means to an end. How could they process this large dataset to accomplish their goal? They'll spend some time learning how to use a distributed system. Some enjoyed this processing and delved even further.

A data scientist can be at various stages of their technical career. They could be at the beginning of their learning to program, have written enough code to complete their thesis, or have worked in or with a software engineer team. Knowing this stage or level of your data scientists is crucial to your success.

Distributed Systems

Some data scientists know nothing about the skills required to work with distributed systems (such as clusters), while others have at best intermediate-level skills. Distributed systems, and their use to create analytical systems, are where things get really complicated—in my estimate, 10–15 times more complex than working with small data.

So data scientists need to know distributed systems enough to accomplish their jobs. This is the area where they'll likely need the most help from the data engineers. The actual level that data scientists need to understand distributed systems depends on how well the data engineers have removed that complexity from the users, and the sheer difficulty of the derivative data product that the data scientists are creating.

Communication

Data scientists will need to verbally communicate their results with the rest of the organization and team. They'll often be tasked with explaining to a business consumer of their model what it means, how they arrived at that conclusion, and whether they think their conclusion is mathematically or statistically correct.

Don't underestimate the importance of this skill. It ultimately determines the uptake and adoption of the data science team's models. Poor verbal communication leads to viewing the models as black boxes or magic. Excellent verbal communication helps people understand the algorithms and reasoning at a high level. Truly effective communication from the data scientist engenders trust in the data products, and that increases adoption.

Visual communication—such as graphs and dashboards—is also a critical part of getting a point across. The data engineering team normally writes the code or creates these graphs and dashboards, but the data scientist should have a sense of will this highlight the important points: for instance, whether to start a scale at zero or focus on a narrow range to highlight data and how to make the key dimension stand out that the organization should track. The data scientist can explain these criteria to the designer or front-end engineer on the data engineering team.

Domain Knowledge

A data scientist can't create models for business problems they don't understand. They won't comprehend it at the level of a business subject matter expert, but they will need to understand as much as possible.

Domain knowledge includes, among other things:

What characterizes the problem?

For instance, are we just trying to reduce defects in some parts of our product? Or is reducing defects part of a larger goal we might achieve some other way?

How appropriate is the data?

Say we have six million rows of customer data—wonderful! But does it really reflect our customers? Maybe the way we collected our data, such as from web visits, excludes older or lower-income customers who are an important part of our clientele.

What other data do we need?

New sources of data are coming online all the time, thanks to improvements in both the hardware and software that collect data. Which of these potential new sources can reveal new insights for us?

The better the domain knowledge, the more likely that the model will solve the business objective or problem. This domain knowledge includes both the business side and the data side. The data scientist will need to understand the data and how it relates to the business domain knowledge.

Technical Debt in Data Science Teams

Data teams working on distributed systems almost inevitably create technical debt: a general concept indicating a certain amount of effort or time the team will have to spend later to rewrite or re-create the system the "right" way. Technical debt is the promise that "We'll go back and do it right later," but all too often, people get busy with new features and never do go back. Technical debt can also come from an amateur implementation or usage that no one realizes is incorrect or suboptimal.

We saw that most data scientists have a novice approach to programming. Along with this comes a difficulty understanding and adhering to the nuances of software engineering. Often, a data scientist writes their code as if it were expressing an equation. That differs from how software engineers write code. Ultimately, data scientists with beginner-level programming skills and no knowledge of software engineering cause a lot of problems. We'll cover these problems throughout this chapter and book.

The software engineering weaknesses that most data scientists suffer from manifest themselves in different ways. Some data scientists are subject to "oneitis"—that is, trying to solve every technical problem with the same technology, no matter what. Data scientists tend not to know or understand what the best technology for each job is. For instance, they will use the same distributed system even if it's really incorrect or suboptimal for their problem. This weakness is also captured by the familiar metaphor about owning a hammer and treating everything as a nail.

During the discovery phase, data scientists need to iterate—and fast. If a data scientist uses the wrong tool, they will vastly hamper their ability to iterate quickly. I've seen data scientists apply the wrong tools and leave themselves the change to run only

20 experiments per day when they should be running thousands. As a result, they'll either have to stop with a suboptimal solution or waste copious amounts of time.

Honestly, this isn't the data scientist's fault. It stems from a management misunderstanding of what a data scientist is and can do. This is really where the data scientists need to interact with the data engineering teams and make sure that the data products are exposed in the ways needed by the data scientists. If you don't have a data engineering team, the management team has no one to blame but themselves.

There's a massive disconnect between what a data scientist thinks they're going to be doing vs. what they actually do. This is made exponentially worse if you don't have a data engineering team at the organization. The data scientist comes in thinking they're going to spend 99 percent of their time doing data science work. The harsh reality is that data scientists spend less time—sometimes far less time—on what they really want to work on: data science.

Google wrote a paper that highlights the disconnect[2] that the vast majority of the time, effort, and problems come from everything around the machine learning model—not the machine learning model itself. From the labels, you can see these are data engineering tasks, or at some organizations, a mix between the two teams.

This hidden focus on sound software engineering is often a surprise to both data scientists and management. In my experience, data science teams are hiding or unaware of the massive amounts of technical debt they've created. This technical debt stems from data scientists with little to no engineering backgrounds creating systems that are difficult for even software engineers—but are within the realm and purview of data engineers and the systems that they should be creating. This technical debt manifests as terrible hacks and workarounds. These hacks make it into production and cause even more problems.

My discussion of a data scientist's journey in the "Programming" section of this chapter repeated the phrase "a means to an end" several times. I emphasized that idea because learning to program or to set up a distributed system wasn't the data scientists' primary focus. Their primary focus was the domain of mathematics. The programming and distributed system skills served as a means for them to accomplish their goal.

There is a significant difference between someone with a cursory knowledge or just enough knowledge and someone with expert knowledge. This is especially true in

---

[2]Read Google's full paper at https://papers.nips.cc/paper/5656-hidden-technical-debt-in-machine-learning-systems.pdf.

distributed systems. If a data scientist puts novice systems or code into production, it leads to significant problems. Companies find this out the hard way.

So your data engineering team is there in part to help the data science team with the distributed systems and programming that they aren't as good at. Going deeper into this subject could overwhelm the book. If you want to find out more about the differences between data scientists and data engineers, check out my in-depth essay on this subject.[3]

Hiring and Training Data Scientists

Since most companies have staff who possess some of the skills required for a data scientist, it's tempting to turn them loose on data science projects. But think carefully before you do this.

The Barriers to Retraining

The incentives for retraining are twofold. First, trained data scientists are rare, and the best ones are snapped up by companies that specialize in analytics. For the same reason, they're also expensive.

The second incentive is that lots of data analysts, database administrators (DBAs), and programmers would love to get their foot in the door of data science. They may see their old roles shrinking and be highly motivated to learn the new skills you need.

So it would seem to be a win-win to upgrade the skills of your existing staff. Unfortunately, I've found that this is hard to do and often fails. The new skills required are just too different from the existing ones that staff tend to know. Depending on the team and their base skills, retraining can be expensive and take a lot of time. Management will need to be careful to weigh the pros and cons of retraining a whole team, individual members of a team, or opt for a different route of outsourcing.

A typical team promotion path to data science is to retrain data analysts or business intelligence staff. This is because the data analysts have the math and statistics background to start off with. On the plus side, these teams should already have an understanding of the business domain and analytics problems that the organization is dealing with. However, not everyone in these teams will be able to make the transition from their previous job title to become a data scientist.

---

[3]www.oreilly.com/content/why-a-data-scientist-is-not-a-data-engineer/

These teams predominantly are missing the programming side and rarely have distributed systems skills. It will take considerable time and effort for people to learn these skills. I've found that a mathematician sees programming very differently from a software engineer or data engineer. Very math-focused people who are learning to program write code that executes math problems. This isn't the greatest or most maintainable code.

Finally, these teams often haven't created machine learning models. The teams will need to be taught the tools and techniques to develop and deploy models. This is also a nontrivial amount of effort.

Some organizations look to their data warehouse or DBA teams for potential data scientists, given their backgrounds in data and analytics. For DBAs, the lack of programming skills (other than SQL) and software engineering skills forms a prohibitively high barrier. And the tools used by data scientists, such as Spark for streaming data and Kubernetes for resource allocation, look nothing like the tools they've used.

For programmers, math and statistics are tall barriers.

So as hard as it may be to explain to your existing staff, data scientists should be hired among those who already possess at least a modestly strong combination of the skills I listed in the "What Skills Are Needed?" section of this chapter.

The worst mistake is to skip the hiring of new resources in order to rely on the existing team. Just renaming a team that doesn't actually qualify as a data science team doesn't fool anyone. It sets your organization and team up for failure.

Upgrading the Skills of Data Scientists

On the other hand, organizations often help their data scientists improve their skills. Generally, this improvement extends to the primary skills that are absolutely needed but often missing or deficient.

To improve their math and machine learning skills, data scientists should read research papers. Many data scientists do this already, but management should realize the value of this and allocate time for data scientists to do it.

Similarly, going to conferences really helps. There are conferences specifically for data scientists or data science topics. These conferences give examples of real-world use of machine learning, while research papers provide a more academic look at what's new. Data scientists should exploit these conferences to learn from others' experiences and find more innovative solutions to their own organization's problems.

Usually, the areas where data scientists are most lacking are programming and distributed systems. Training them in these areas can provide the most significant improvement in their productivity, but also lead to the risk of diminishing returns. There is definitely a happy medium to this improvement.

Put a different way, if a data scientist is an absolute beginner at distributed systems, there is value—probably significant—to improving their knowledge there. It helps the data scientist to be more self-supportive and leave the data engineering team alone. On the other hand, if a data scientist is already pretty skilled with distributed systems, putting more time and effort into learning a distributed system may not have a payoff. Further training might just teach them advanced concepts or use cases that they're never going to implement or have the ability to execute.

Finding Data Scientists

Most companies will have to recruit data scientists, either to establish a new team or to increase the size of an existing team.

There aren't many programs for data scientists in universities. And often, these university programs are limited in scope and are academically focused instead of industry-focused, so they don't reflect the real day-to-day life of a data scientist.

There are also commercial and boot camp programs. The people coming out of these programs vary in their ability to be productive on a team. As you hire, keep in mind that even if you find someone, you may need to teach them a lot and be patient.

So most of these programs can't cover everything that is really required for the day-to-day work of a data scientist. Hiring from these programs is more of an investment, where you trust that the person you've chosen can grow and has the potential to get to the next level.

Recruiting outside and inside talent will be difficult. There is a great deal of demand for data scientists of all levels of experience. Make sure that you start recruiting long before you actually need the person. You may need a 6- to 12-month lead time to hire someone. Often the candidates will have multiple offers, and extending an offer doesn't mean the person will join the organization.

Remember that not everyone who is a possible data scientist has that title, has a degree in math, or comes from a specific math-focused background. Some of these data scientists will come from the physical sciences world.

Many of my clients and other companies have had success with physical science majors. They're people who had to learn to program and some distributed systems to do their professional work or academic thesis. Much of this work dealt with getting data, analyzing data, and making decisions based on data. This is precisely the sort of work and data thinking that data scientists do.

Meeting the Needs of Data Scientists

Managing data scientists brings some unique and challenging changes to data teams. This primarily stems from a discipline that is comparatively new and doesn't entirely fall into one slot or another. In some ways, data science falls into its own slot but still borrows ideas and techniques from other disciplines. This borrowing makes managing a data science team difficult. Here are a few things to know:

- The majority of data scientists have never written big programs, measured either in lines of code or in systemic complexity.

- Data scientists may have never worked in an enterprise or big organization and don't know all of the issues that come along with working in a big organization.

- Many data scientists come from an academic background, either having just completed a Ph.D. or having worked in academia, and this is their first nonacademic job.

- Many data scientists have taught themselves coding but have never worked in or with a professional engineering software team.

- Some data scientists will overestimate their abilities in engineering and programming.

- Some data scientists will underestimate the difficulty of data engineering, especially concerning the complexity associated with distributed systems.

- Some data scientists have never dealt with the messy and imperfect datasets of the real world.

Knowing that data scientists don't fit into an organization's precreated buckets, the management for data teams will need to make allowances and considerations.

Introducing Software Engineering Practices

Sometimes, data scientists mistakenly consider themselves programmers or data engineers. After all, the data scientists do create software. Although occasionally a data scientist learns the software engineering rules and best practices that come along with creating code, most data scientists don't understand what's needed.

Some engineering best practices are bare minimums, while others are good to have but dispensable. To pick an example, I consider source control to be a bare minimum for a data science team. You need source control to track how code changed when you introduce an error, to offer a single safe place for source code, to enable code review, and to allow multiple people to work on the code with minimal friction.

On the other hand, I consider continuous integration and continuous deployment, just a nice process to have. Some data science teams have this in place, and it definitely helps. It's also something that a majority of data scientists don't know how to set up or operate. This would be an area where the data engineering team can be useful to set up the process and show the data scientists how to use it.

Generally, software engineers and data engineers love rules and order. Data scientists feel that data engineers tend to overcomplicate and overengineer things. This perception makes data scientists push back on what they perceive to be rules that slow them down and prevent experimentation. There really is a quantifiable difference in personality between data scientists and data engineers. Because of this difference, management needs to know that they will have to pick their battles concerning which best practices to require. Enforcing every best practice will lead the data scientists to push back.

Too Much Process Stifles Progress

It's difficult to put something like data science into a single box, because it really straddles disciplines. It's somewhere between science and software engineering. Furthermore, data science is an inherently creative process. Imposing too much of a heavyweight process will inhibit the creativity of the data scientists. On the other hand, a lack of process could make the data science team an unaccountable free for all.

The key for management is to find that happy medium with just enough process in place. They'll have to figure out which methods are enough to keep chaos at bay. Remember that not every problem that comes up needs a heavyweight rule to prevent it from ever happening again.

Sometimes managers ask one team to copy processes from other teams or decree an organization-wide mandate to use a particular method. These sorts of rigorous software engineering processes may work well for data engineering, but not as well for the data science team. It's essential to understand how data science projects are different from software engineering.

Engineering processes look fondly on precise outcomes that always result in a specific artifact, be that a whole new product, a bug fix, or a new product feature. But data science projects are open-ended. There are several possible outcomes for a data science project, which are more like research than engineering. As a result, the findings mirror those of research too. An entirely feasible outcome for a data science project is to disprove the hypothesis. In this situation, there is no actual artifact or deliverable, and the data scientists themselves didn't fail. They simply proved that a proposal won't work.

As another example of a problem, some organizations try to force a Scrum project management framework on their data scientists. This requires a data scientist to use a framework that wasn't designed to handle their sort of work.

In most cases, managers just want to know the status of a project or what the data scientist is working on. Organizations can often thrive by decreeing only enough processes to understand and report on outcomes or status.

Striking this balance may entail educating other managers and parts of the organization about the nature of data science. They may be expecting data science to function closer to a business intelligence function, where data comes in and reports go out. They could expect data science to be like a software engineering team, where software is worked on consistently and (at least in theory) with consistent software releases. Instead, organizations may need to understand that the data science team works more like research, with time and data going in and an unknown result coming out within an unknown time frame. To combat research projects that could go on forever, some organizations will timebox their research. With a timebox, the researcher will move onto something else after a certain amount of unproductive time.

CHAPTER 4

The Data Engineering Team

Looking for a man with a focus and a temper
Who can open up a map and see between one and two

—"Teen Age Riot" by Sonic Youth

A data engineering team is responsible for creating data products and the architecture to build data products. These data products are really the lifeblood of the rest of the organization. The rest of the organization either consumes these data products—deriving insights that drive planning—or creates derivative data products for further use.

On the architecture side, it is up to the data engineering team to create data pipelines. A data pipeline is a process to take raw data and transform it so that it becomes usable by the entire organization. To accomplish this task, the data engineering team must choose the right technologies for the data and the use cases: for instance, which of the many available data stores to place data in, which message queueing system to pass data through, and so on.

The data engineer must also maintain the data in formats ready for use, keeping in mind the tendency for both the shape of data and the enterprise demand for it to change.[1] Data pipelines also need systems to run on, either on-premises and in the cloud, and these systems have to scale widely.

Data pipelines are coded using the APIs of the various tools they invoke. So writing and testing that code is also the data engineering team's job. This really puts the onus on the data engineer to master the necessary programming and to understand the distributed systems they deal with. The data engineering team is responsible

---

[1]Read my and other viewpoints on data pipelines in my blog posting, What Is a Data Pipeline? (www.jesse-anderson.com/2018/08/what-is-a-data-pipeline/).

J. Anderson, *Data Teams*, https://doi.org/10.1007/978-1-4842-6228-3_4

for validating the architecture they create and for ensuring users that it will scale appropriately. Therefore, robust software engineering techniques such as continuous integration and continuous delivery (CI/CD) are also part of their skill set. Without a robust architecture and data infrastructure, the project will underperform or go nowhere.

The team isn't just made up of data engineers. Unlike the data science team in the previous chapter, the data engineering team is cross-functional, and you will have other job titles on the team. Data engineers will predominate, but other people, such as front-end engineers or graphic designers, will complement them.

Thus, a data engineer is a software engineer with specialized skills in creating scalable, production-hardened solutions around big data.

In this chapter, we'll look at what a data engineer is and what skills they need. We'll also talk more about the multidisciplinary composition of this team and give advice about what to do when you need to bring existing nondata engineering staff onto a data engineering team. We'll also clear up misconceptions about the relationships between data engineering and data warehousing and between data engineering and data wrangling.

What Skills Are Needed?

A software engineering background is highly recommended for a data engineer. Software engineering is a foundational discipline for this role, and someone who hasn't been trained in the design of large software systems will probably create fragile and hard-to-maintain solutions.

Data engineers have further specialized their software engineering to apply them to big data and distributed systems. During their small data and software engineering days, they may have started doing extensive multithreading or client/server systems. This background is beneficial when a data engineer begins to work on distributed systems. They aren't 100 percent the same, but the experience serves as a foundation for learning.

Data engineers also have an interest in data itself. Some would go so far as to say they have a love of data—and I would agree. In their past, these software engineers may have taken it upon themselves to create analytics or put effort into capturing data that your average software engineer wouldn't. They realize the importance and value of data.

Data engineering teams tend to skew toward senior people. That isn't to say that data engineering teams never have junior people. It's more that data engineers have come through the software engineer ranks, so teams should include people who have been mid-level or senior-level software engineers. Junior members of a data engineering team often have a Master's or advanced degree, or they've previously specialized in some form of distributed systems or big data projects.

More often than software engineering, data engineers encounter inevitable trade-offs that come with scale. For instance, a hard and fast rule of software engineering says to store all data in a single place (normalized) with a single technology. Most commonly, software engineers are storing their data in a relational database in a completely normalized way. When creating a system at a large scale, the data is already joined together (denormalized) and stored in several different technologies that are meant for particular use cases.

A software engineer may feel the need to adhere to best practices that are well established at a small scale, but that doesn't work well on a large scale. A data engineer will need to understand when to adhere to the best practices and when it makes sense to deviate in pursuit of speed or to deliver higher scalability.

Relational databases and data warehouses will remain part of the organization's datasets for the foreseeable future. So some conventional SQL skills should be part of the data engineer's toolkit as well.

The following subsections go into detail about the following skills, which are needed on a data engineering team:

- Distributed systems

- Programming

- Analysis

- Visual communication

- Verbal communication

- SQL

- Schema

- Domain knowledge

- Other important skills

While the majority of the data engineering team needs to have strong programming and distributed systems skills, the other skills don't need to be embodied by every single person on the team. Depending on the use cases and data products, these additional skills can be incorporated by one person or a few people on the team.

Distributed Systems

Distributed systems are hard. You're taking many different computers and making them work together. This requires systems to be designed differently. You have to plan how data moves around those computers.

Having taught distributed systems for many years, I know this is something that takes people time to understand and get right.

Data engineers will need intermediate-level to advanced-level abilities to create and use distributed systems. A few such skills include an understanding of resource allocation and of network bandwidth requirements, how to create virtual machines and containers, replication, how to partition datasets and message queues, and how to handle failures and fault tolerance.

Programming

The data engineers on your data engineering team are tasked with writing the code that executes the use case on the big data framework, so they need to be skilled programmers.[2]

The actual code for big data frameworks isn't tricky. Usually, the most significant difficulty is keeping all of the different technologies straight. With small data, programmers usually need to know only one to three technologies. The programmers on data engineering teams will need to know 10 to 30 different distributed systems or data technologies: the APIs, architectures, and application of each technology.

By programming, I don't mean just a knowledge of syntax. The data engineers are also responsible for continuous integration, unit tests, and engineering processes. These needs are often misunderstood and left missing from teams. Sometimes data

---

[2]In this case, a programming language means a procedural language such as Java, Python, or Scala. Data engineers should know SQL in addition to a procedural language. Data engineers or other SQL-focused people that only know SQL will need to learn a procedural language and that will be a difficult and time-consuming task.

engineering teams forget that they are still doing software engineering and operate as if they've forgotten their software engineering fundamentals. These requirements don't go away—in fact, they're more important than ever due to the sheer complexity of big data problems.

Before a data engineer fixes an issue, they should write a unit or integration test that simulates the problem or error state. This unit test should fail at first, confirming the existence of a problem, and then pass once the data engineer has fixed the issue. Without extensive unit testing or integration tests, the data engineering team won't know what other issues could have been created by fixing the first issue.

Analysis

A data engineering team will need to expose some data analysis as a data product. Analytical skills enable the team to create these analytic data products. This analysis can range from simple counts and sums to more complex products that extract new dimensions from data.

The actual skill level can vary dramatically on data engineering teams; it will depend entirely on the use case and organization. The quickest way to judge the skill level needed for the analysis is to look at the complexity of the data products. Are they relatively straightforward, or do they involve equations that most programmers wouldn't understand?

Sometimes, an analytic data product is a simple report that's given to another business unit. This could be done with something as simple as SQL queries. More advanced analytics will come from the data science team.

Visual Communication

A data engineering team needs to communicate its data products visually. This is often the best way to show what's happening with data—especially when there are extraordinarily vast amounts of it—so that others can readily use the results. You'll usually have to demonstrate data over time and with animation. This function combines programming and visualization.

A team member with visual communication skills will help you tell a graphic story with your data. They can show the data not just in a logical way, but with the right aesthetics.

Most software and data engineers aren't known for their graphical abilities and beautiful user interfaces. You can spend copious hours and resources trying to get a data engineer to improve their artistic skills. Or you could place a front-end engineer on the data engineering team to make the customer-facing data look and act correctly.

How do you handle the lack of understanding that the front-end engineer will probably suffer from when it comes to the data and how it was produced? Make the front-end engineer work with the data engineer to comprehend what they're working on. The data engineer could write code that pulls the data in a way that the front-end engineer can start working with. Another possibility is for the two people to pair program for a time.

As you saw in the "Communication" section in Chapter 3, the data science team also needs to employ visual communication. Depending on the amount of front-end work that needs to be done, the front-end engineer's time could be split between the data science and data engineering team. Alternatively, all front-end work for data could be the sole purview of the data engineering team. In that case, this team would assign a predefined portion of front-end data work to the front-end engineer.

Verbal Communication

The data engineering team is the hub where many spokes of the organization enter. You need people on the team who can communicate verbally with the other parts of your organization. This skill is usually fulfilled by a data engineer or architect because there needs to be a deep understanding of the technical issues, the data, and the users' interests and preferred terminology.

Your verbal communicator is responsible for helping other teams be successful in using the big data platform or data products. They'll also need to convey to these teams what data is available. Some data engineering teams operate like internal solutions consultants.

This skill can mean the difference between increasing internal usage of the cluster and seeing the work go to waste. Without an internal evangelist or someone who can speak coherently about the data products, the data may never get used.

SQL

Rounding out a team with a database administrator (DBA) helps the team fulfill its need for the SQL skill. An SQL expert can also help at a much later stage in the data engineering life cycle. Once the data and data infrastructure is in a place, most if not all data is exposed with an SQL interface. At that point, the SQL-focused people can start providing value. This also accounts for situations where others, usually business teams, are unable to write the SQL queries themselves.

Here are some titles that I classify as SQL-focused:

- DBA (Database Administrator)
- Data Warehouse Engineer
- ETL Developer
- SQL Developer
- (Some ETL Technology) Developer

Throughout the book, you may see me generally say use the term DBA to mean this list of titles instead of listing these titles explicitly.

Of course, a data engineering team can't consist only of DBAs or SQL-focused people. If a team made up of only DBAs like this actually gets something into production, it's more of an outlier than a real success. In dealing with these sorts of teams, I've found that their "production" system is held together with more duct tape and hope than anything else. These systems break all the time and are virtually impossible to upgrade with even the smallest improvement.

Conversely, a data engineering team that has no DBAs will lead to a different type of failure. There are efforts and pieces of the data puzzle that DBAs have spent their entire career dealing with.[3]

---

[3]I've found DBA's experience especially helpful in highly regulated environments where the same regulatory and compliance issues have to be kept, no matter what the scale of the data that's being processed. These DBAs can share the regulations and how other systems conformed to the rules.

Schema

Although some big data tools and storage systems are advertised as "schemaless," they always embody some kind of key/value structure or tags. They effectively have schemas, even though the schemas might not be as rigid and predictable as relational data schemas.

Still, unlike most software engineering, big data is often not exposed through an API such as a REST call to the organization. Instead, the data is often exposed to the rest of the data teams—or the organization—as the raw data. A schema makes sure this data is presented correctly.

Thus, members with this skill help teams lay out data. They're responsible for creating the data definitions and designing its representation when it is stored, retrieved, transmitted, or received.

Despite the constant need to define and apply schemas, this skill is often missing from data engineering teams. The importance of this skill really emerges as data pipelines mature. I tell my clients that this is the skill that makes or breaks you as your data pipelines become more complex. When you have a petabyte of data saved on HDFS or blob storage, you can't rewrite it each time a new field is added. Skill with schemas helps you look at the data you have and the data you need and then define what your data looks like.

When working with teams, I repeatedly find questions to ask about schemas. In general, the DBAs can answer the problem six times as often as their software engineering counterparts. DBAs often bring the schema skill to a data engineering team.

Often, teams will choose a simple but inefficient data format such as JSON or XML. As a result, 25 percent to 50 percent of the information is just the overhead of tags. It also means that data has to be serialized and deserialized every time it needs to be used. The schema skill on the team will help choose and explain the right way to expose data. They also advocate for binary formats like Avro or Protobuf. They know they need to do this because data usage grows as other groups in a company hear about its existence and capabilities. A format like Avro will keep the data engineering team from having to type-check everyone's code for correctness.

The schema skill goes beyond simple data modeling. Practitioners should understand the basic computer science behind storage choices, such as the difference between saving data as a string and a binary integer.

Domain Knowledge

Some jobs and companies aren't technology-focused; they're actually domain expertise-focused. These jobs focus 80 percent of their effort on understanding and implementing the domain. They target the other 20 percent on getting the technology right. Domain-focused jobs are especially prevalent in finance, health care, consumer products, and other similar companies.

Domain knowledge needs to be part of the data engineering team. As many people on the team as possible should have this background knowledge. These people will need to know how the whole system works throughout the entire company. They'll need to deeply understand the domain for which you're creating data products; these data products will need to reflect this domain so they can be used within it.

Other Important Skills

Some skills are less often recognized and may be lacking from data engineering teams. People who came up through the DBA ranks are more likely than those from software engineering to understand the importance of these requirements. DBAs have spent their careers on such things, whereas most software engineers have never dealt with them.

Software engineers may not immediately realize the value of these skills because they've never been bitten or experienced the sheer need for these requirements. It comes back to the issue that a data engineering team without a DBA will have a late-stage failure that's a complete blind spot. I expect the DBAs to be the ones on the data engineering team that are screaming and making a case for these requirements.

These requirements also tend to be recognized only by those who have worked for larger organizations, because large organizations have to fulfill these requirements on the majority of their projects.

Here are the more important ones that are often missing:

Data Governance

How secure is the data? Should the data be masked? Does it have personally identifiable information (PII) that needs to be hidden?

Data Lineage

Where did the data originally come from? When there is a problem with the data, where do we look back to find the source data?

Data Metadata

What happens when you have thousands of data products? How do you keep track of their metadata and schemas?

Discovery Systems

How do you catalog your datasets and help potential users find them?

BIG DATA ENGINEER AND OTHER JOB DESCRIPTIONS

Some organizations use different job titles for data engineers. The most common is "big data engineer." But I've found that the title "data engineer" is the most common, so I recommend you advertise for that job title to cast the widest net possible.

If you're thinking of hiring someone whose previous job had a different title, make sure that person has done the same functions as a data engineer.

To add to the confusion, the title "data engineer" is used in some organizations for a very different job function with a different skill set focused on relational data and SQL. That role is in no way preparation for being a data engineer in the way defined in this book.

Organizations that fail to understand the difference between an SQL focus and a software engineering focus will not be successful in their data products. Data engineering teams really do require programming and distributed systems skills. SQL-focused data engineers lack both of these crucial skills.

Levels of Expertise

It's useful to have three levels of expertise in data engineering. At the lowest level are people supporting the projects day to day. These may be new to the field of data engineering. Once someone has carried through a major data engineering project successfully, they can be considered a *qualified data engineer* with greater responsibilities. And the highest level is called a *veteran* in this book. The following sections lay out the responsibilities of these two higher-level positions.

New Data Engineers

A person who is new to the field of data engineering is a great accomplishment. They've put the time and effort into specializing their skills even further. They're eager to learn and start getting experience.

Organizations should know that some of the ideas that need the most critiques and help come from their newest hires. New data engineers often know "what" but not "why." That missing "why" can lead to significant rewrites of code. Sometimes, new data engineers will get their "what" from a vendor's white paper or marketing material. A new data engineer won't know the difference between what really works and what doesn't. They may haul off and invest months of work in implementing something that isn't a viable solution.

Qualified Data Engineer

A qualified data engineer has actually put a system into production. I highly recommend having at least one qualified data engineer on a data engineering team, because much of what a data engineer knows must be based in experience. They have to answer questions such as: Should you use one technology or another? Should you use one architectural pattern or another?

The only way to get this knowledge is to actually get your hands dirty in data engineering, put something in production, gotten, and been bitten by any problems that came up in production.

Veteran

Going beyond the qualified data engineer, a veteran has far more experience in putting distributed systems into production. A veteran should have put many projects into production and therefore wrestled with many different issues and have the battle scars to prove it. Putting a single distributed system into production gives some experience, but not enough to be a veteran. Data engineering teams should have both veterans and qualified data engineers on them.

53

The veteran should have extensive experience in distributed systems, or at least extensive multithreading experience. This person brings a great deal of experience to the team.

The veteran is crucial for holding the team back from bad ideas. The veteran has the experience to realize when something that's technically feasible is a bad idea in the real world. They will give the team some footing or long-term viewpoint on distributed systems. This translates into a better design that saves money and time once things are in production.[4]

Further Specialization

Due to the sheer complexity and growing number of distributed systems technologies, data engineers may need to specialize further. By specializing, the data engineers will enjoy a higher degree of knowledge and can optimize the creation of data products.

It's common for data engineers to run the infrastructure in the cloud. As cloud providers mature, they are creating more and more specialized products for building data products. These specialized products could be managed services created by the cloud provider, or open source tools with the cloud provider managing the infrastructure behind the scenes. Some managed services will use an open source API, and others will use a custom API. When using the technology, there may be additional caveats that don't exist in the original open source product. In any case, the data engineer may need to specialize in a specific cloud provider's line of products to keep up with all of the changes and caveats.

Distributed databases offer a plethora of choices. Examined in a cursory manner, many databases may seem to offer the same features and be good for the same use cases. However, the core differences and caveats may mean the difference between success and failure for a project because one particular database may not be able to handle a use case correctly. Rewrites and rearchitecting a data product or solution can be incredibly expensive. This is a big reason we'll see more data engineers start to specialize further into databases.

---

[4]Read more about the importance of project veterans in my blog posting The Veteran Skill on a Data Engineering Team (`www.jesse-anderson.com/2018/02/the-veteran-skill-on-a-data-engineering-team/`).

Another specialization for data engineers is creating real-time or streaming systems. At a superficial level, real-time systems seem like the simple addition of a new real-time or streaming technology. The reality is that the real-time constraints add a much deeper level of complexity. Real-time restrictions require a new level of understanding about topics that don't exist in batch systems, such as exactly once execution, idempotent systems, and error handling. I expect real-time systems to become a more prevalent specialization with data engineers.

Should the Data Engineering Team Focus Only on Big Data?

A common question concerns the scale of data processed by the data engineering team. Should this focus only be on big data? Or can it include small data?

A data engineer should be able to handle both big data and small data. This principle stems from the career progression of a data engineer. They should have mastered the small data side of things before they came over to big data. Data engineers should have a pragmatic view of data to choose the best tool for the job.

Given that a data engineer can do small data, should they? Should you drive your Ferrari to get groceries? Maybe. The Ferrari will get you there, no doubt. Along the way, you face some risks that wouldn't have if you'd driven your other car. There's the risk that the Ferrari will get hurt in the parking lot or—worse yet—get stolen. Likewise, on a data engineering team, you've put all this time and money into hiring or training someone with a specialization. It's probably a better use of their time to apply their specialty where it's most needed and have someone else without that specialization do that small-scale work.

If you're a small business or startup, you may not have enough people to afford a real specialization in big data. At these smaller places, data engineers may have to go back and forth by working on projects of different scales. This constant back and forth movement can cause major context switches that can take data engineers a while to make. For larger organizations, a split between staff, with particular specializations, makes sense. These larger organizations give you the ability to designate teams or individuals for specific data scales.

Another thing to keep in mind is that the scale isn't always big data. Sometimes it's in that in-between world of medium data. Not everyone has truly big data. I think medium data is far more prevalent, so I coined that term.[5] Just remember that everything outside of small data needs distributed systems—this could be medium or big data. Either way, you'll need a data engineer to handle it.

Common Misconception

Although readers who have started at the beginning will understand the key distinctions made in this book, certain misunderstandings come up so often in my work that I want to explicitly address them here.

Why Data Engineering Isn't Just Data Warehousing

A common misconception is that data engineering is simply a new name for data warehousing. This misconception has led to the failure of many projects because the data engineering and data warehousing teams require different skills and compositions. This difference ensures that locating a big data project in a data warehousing team has very low odds of success.

Data warehousing teams are mostly SQL-focused. They lack the necessary programming and distributed systems skills that data engineering teams require. The technologies used by data warehouse teams are primarily mature and prewritten technologies. The team didn't need to write any code because the data warehouse technology vendor already wrote all of the code they needed.

Data engineering is an entirely different animal. The team has to bring together many different technologies to make a data pipeline. These technologies tend to be open source technologies that work on distributed systems. The technologies themselves tend to have pointy edges and aren't as mature. More importantly, they need code to be written, and you can't just use SQL for everything.[6]

---

[5]`www.jesse-anderson.com/2017/06/medium-data/`

[6]Read my blog post to see why you can't do everything with SQL (`www.jesse-anderson.com/2018/10/why-you-cant-do-all-of-your-data-engineering-with-sql/`).

Why a Data Engineer Isn't a Data Scientist

There is a great deal of confusion about the difference between data engineers and data scientists. This confusion comes from visualizations like Figure 4-1. They really overgeneralize the relationship and overlap between the two titles.

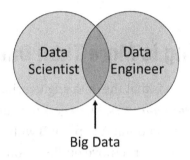

Figure 4-1. *A poor visualization of the differences between data engineers and data scientists*

As you've seen, the reality is far more complicated. Figure 4-2 shows these nuances better.

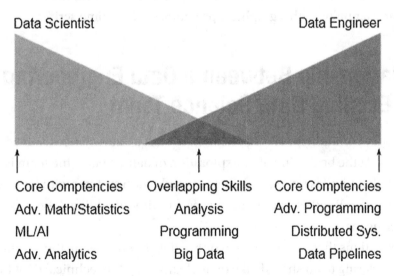

Figure 4-2. *A better visualization of the differences and strong suits between data engineers and data scientists*

A data engineer is rarely mistaken for a data scientist. Management doesn't expect a data engineer to create a model. It's far more common for management to mistake a data scientist for a data engineer. Management will expect a data scientist to do the tasks and roles of a data engineer.[7] There are some "full-stack" data scientists out there, but very few and far between. Data engineers are far better at programming and distributed systems than data scientists.

Why Data Engineering Is More Than Data Wrangling

Another common misconception is that the data engineering team is a collection of data wranglers, sometimes called *data mungers*. Managers may think that data engineering is a simple task of copying data from point A to point B with some simple data transformations to harmonize fields or remove outliers along the way. In this scenario, often, they will move the data from one "raw" data format into another format.

This understanding puts data engineering on the very low end of what should be accomplished by the data engineering team. I would argue this bar is so low that it doesn't even keep to the definition of a data engineer. Data wranglers lack both the programming and distributed systems skills required of a data engineer. Most of their tasks are done manually, with a graphical program, or with some scripts.

The Relationship Between a Data Engineering Team and an Existing Data Science Team

It's unfortunately common for an organization to start a data science team without data engineers. At the beginning of its exploration of data science, the team is in an experimental phase and is not responsible for production systems with long-term maintenance needs. But they cannot be successful without adding the robustness of a data engineering team.

For organizations that previously had only a data science team, the first task that the new data engineering team should take on is cleaning up the technical debt from the

---

[7]I've written extensively on this subject because it is such a common mistake and misconception. I highly recommend you read "Data engineers vs. data scientists" (`www.oreilly.com/ideas/data-engineers-vs-data-scientists`) and "Why a data scientist is not a data engineer" (`www.oreilly.com/ideas/why-a-data-scientist-is-not-a-data-engineer`).

data science team. (See the "Technical Debt in Data Science Teams" section in Chapter 3 for the reasons technical debt emerges.) The debt exists because data scientists are often beginner programmers. They haven't written large code bases and don't know the issues of maintaining and dealing with them. Data engineers do—or at least should—know how to manage large code bases and the proper distributed systems to run them on.

The actual technical debt can be many and varied. The entire system might be held together with duct tape and hope, or there could be short-term trade-offs in the code itself that hold back further development. In either case, management should get an estimate of the amount of time and effort it will take to pay down or rewrite the data scientist team's code.

Retraining Existing Staff

Some companies will want or need to look internally for data engineers. More than likely, these people won't have all of the skills to be a data engineer from the beginning. It is incumbent upon the organization to give these new data engineers the resources to succeed. These resources include learning new programming languages, new distributed systems, and new ways of creating data pipelines. When an organization doesn't provide these resources, they are setting themselves, the team, and the individual team members up for failure.

When doing internal promotion, look for people with a background in multithreading, multiprocessing, or client/server systems. This will give them some basic understanding of how distributed systems work and can be used. From there, I look for a desire to learn and expand their technical stack. Even with a software engineering background, learning distributed systems required significant time and effort. Often, these people have already taken steps to start learning the systems, or have shown a willingness to get started.

On top of all this, they need to show a love of data. This manifests as having an interest in processing data and value creation from data. They may have already coded some analytics or data pipelines on a small scale.

Most of the time, when an organization looks internally, they assume they can ameliorate an individual's skills gap. But you should also be on the lookout for an *ability gap*. A skills gap means that a person simply lacks the knowledge, but has the innate ability to be successful with distributed systems. An ability gap means that a person lacks both the knowledge and the innate traits that will allow them to acquire the skills. No amount of time, resources, or expert help will get such unfortunate staff to that next level.[8]

In the "Qualified Data Engineer" section earlier in this chapter, I advised hiring at least one qualified data engineer who has all the experience required to create a production system. When you promote existing staff, they more than likely won't serve as qualified data engineers. For some organizations, it becomes absolutely necessary to turn a nondata engineer from the inside into a data engineer. If so, it is incumbent on you to give them the additional mentoring and project veteran resources to eventually become a qualified data engineer and even a project veteran. If the company lacks internal resources for mentoring and project veterans, they may need to look outside the organization to help their internal resources improve.

Software Engineers

When an organization starts to look internally for potential data engineers, I recommend starting with their software engineers. This is generally because the software development skills are the most difficult to obtain, out of the listed in "What Skills Are Needed?" section earlier in this chapter.

Remember that in contrast to the data science team, where the technical bar is much lower, data engineers need a much, much higher technical programming level to be successful. Even with a software engineering background, it still takes time to feel comfortable with the vast ecosystem of distributed systems. In my experience, an excellent software engineer will feel comfortable doing data engineering after three months of study. An average software engineer will take six to twelve months. Management needs to take this into consideration when making decisions about project timings and hiring.

Management must also realize that not every software engineer can make the switch to data engineering, although they enjoy higher odds than people in other disciplines. Management needs to be judicious in vetting their potential data engineers.

---

[8]Yes, this ability gap is harsh and sounds elitist, but it comes from direct experience. I coined the term after spending an extensive amount of time teaching people from all backgrounds and experience levels. I started to see patterns in peoples' inability to understand even the basics. I coined the term "ability gap" to let management know about this issue. Read more at Ability Gap—Why We Need Data Engineers (`www.jesse-anderson.com/2016/07/ability-gap-why-we-need-data-engineers/`).

SQL-Focused Positions

Often, management will look to their SQL-focused positions as a source of data engineers. Their thinking is that data engineering is just some basic ETL (extract, transform, load) workflows and doesn't really require programming or distributed systems skills. Their logic is that these people have become expert at handling the organization's data at small scales, so they can make the jump to a more massive scale. This is a mindset that managers should avoid.

As explained earlier, data engineering teams made up of only SQL-focused people will generate little value relative to investment—so low that I advise teams not even to attempt the projects.

Having taught—or tried to teach—a significant number of SQL-focused people distributed systems, I can say this will be a very, very difficult journey. There are two main gaps in the skills of SQL-focused people, so far as data engineering is concerned: programming and distributed systems. In other words, they're lacking in the two main areas where data engineers need to be more than just OK—they need to excel.

I've seen an incorrect perception SQL-focused people can get by using a high-level tool that doesn't require programming. But that leaves the far from a trivial matter of understanding distributed systems. The tools don't negate the need for a data engineer to understand the distributed systems they supposed to be using, architecting for, and debugging. Even with these tools, the value creation is so limited or low that it isn't worth attempting the project.

All of this said, where does an SQL-focused person fit? It is possible that SQL-focused people can become data engineers, but not as likely. As time goes by on the project, the data engineering team will have had time to polish how they expose data to the rest of the organization, usually with a technology that has an SQL interface for accessing data. These technologies lower the technical bar to the point where an SQL-focused person can do some tasks. These could be tasks such as creating a report based on an SQL query or helping other parts of the organization with writing an SQL query for an analytic. This maturity comes at a much later stage in the projects, and it's at this point that the data can be exposed to SQL-focused people with SQL interfaces. However, starting a data engineering project with SQL-focused people is fraught with failure.

The Role of Architects

Architects are senior or principal programmers who have taken on the higher-level task for viewing a system as a whole and ensuring that it has business value. The architects are the glue between the business and its technology. An architect is expected to make the tough and long-term choices about which technologies to use. They verify that the proposed architecture meets the business needs.

In data engineering, architects map out the interactions in a pipeline and diagram out those interactions. The interactions are quite complex and therefore require a programming background, not just a knowledge of SQL and schemas.

To be honest, data architects who lack a programming background often disagree with that assessment. But here's the clincher: data architects who started an SQL background and then learned how to program agree with me.

So, what skills do architects of big data pipelines need? Having consulted with many different types of architects over the years, I'll offer some general suggestions here.

I believe architects for data engineering teams need to know how to code and should have been a software engineer at some point in their careers. They may not code as much anymore, but they should have coded at some point. To have experience in distributed systems, they should have ideally coded or architected a distributed system before. There really isn't a substitute for experience in this area. Really, the way to have gained this experience is firsthand in the dirty and grimy trenches of doing it.

Architects need to understand the sheer difficulty of what they're proposing. Creating an architecture with distributed systems isn't about arbitrarily choosing technologies and drawing boxes. The architect must deeply understand the use cases and how the data will be accessed. With this knowledge, they can use their in-depth understanding of the technologies themselves to choose the right tool for the job. Architects who blindly follow vendor recommendations, white papers, or whatever is the hottest technology will fail to create the right architecture.

Architects who didn't code and came through more of a DBA/SQL-focused route often have a difficult time understanding the distributed systems they are tasked with architecting. They may understand the basic concepts, but they have difficulty understanding the technology to the level that an architect should in order to choose the right tool for the job.

To add to the difficulty of choosing a technical path, some technologies just aren't mature enough for production use. The architectural patterns aren't very well defined either. In the future, architectural choices will become easier for very specific use cases or verticals. This stands in contrast to general-purpose use cases where most data engineering teams have historically had to write custom code. I don't think these for general-purpose computing projects will become easier or purchasable as part of an off-the-shelf product.

There are subtle nuances and tricky trade-offs in distributed systems. I call these cheats. To achieve the necessary scale, distributed systems have to cheat in one way or another. Sometimes people call these choice "trade-offs," but that term makes it sound like a small thing. The reality is that the choice of trade-offs in big data can make a use case possible or impossible.[9] Architects must possess the knowledge and understanding of the effects of these cheats on architecture and how they interact. This is indeed what separates the mediocre architects from the great architects. A holistic approach, in turn, makes data projects successful.

Because the data engineering team skew toward more senior people, architects may get more push back on their designs from the data engineering team. The architect needs to have gained the trust and respect of the team. This will make it easier for the architect to make more unorthodox or long-term calls on technology paths. If at all possible, engage the rest of the data engineering team on technology choices. It is far easier to get the team's buy-in when they feel like they're part of the decision-making process instead of being dictated to.

Placement in the Organization

On what team should you put your architects? Should they be part of the data engineering team or part of a broader architecture group?

I've seen both choices produce success. I've seen the architect be part of a separate architecture team, and I've seen them be a member of the data engineering team. The choice really depends on the size of the organization and how the data engineering team came to be formed.

If the architect is not part of the data engineering team, there are several critical factors to success. First, the architect needs to have a very high-bandwidth connection to the data engineering team. The architect's decisions and technology choices need to be communicated to the data engineering team, who must reliably follow through on them. Any perception that an architect is disconnected and aloof from the team will

---

[9]Read more about what I mean at On Cheating with Big Data (www.jesse-anderson.com/2017/10/on-cheating-with-big-data/).

cause friction. Second, the architect needs to realize that distributed systems require a deep specialization that isn't possible with a cursory look or by copying a vendor's architecture white paper. The architect needs to deeply understand the various nuances of data engineering and keep up with the constant changes in the landscape.

If you have the choice, I recommend having the architect be part of the data engineering team. Data engineering is really a specialty and keeping everyone together really helps.

In some organizations, there isn't a specific or single person that has the role of architect. Instead, the data engineering team as a group fills the architect role. They are responsible for both the software architecture and the coding of the system. No one person on the team has the specific title of architect. Instead, the entire data engineering team works together to create software architecture.

CHAPTER 5

The Operations Team

No need to ask
He's a smooth operator

—"Smooth Operator" by Sade

An operations team is responsible for putting both cluster software and custom software into production and making sure it runs smoothly. They are the people who guaranteed customer access to the service. Without this team, the software can't run consistently enough for end users to trust it. If the system is continually failing in production, you will lose users and forgo adoption because the system is too unstable. A good operations team prevents this from happening.

The operations team is primarily made up of operations engineers. An operations engineer has an operational or systems engineering background along with specialized skills in big data operations and understands data and some programming or scripting. These staff members have extensive knowledge of hardware, software, and operating systems. They use this knowledge to keep the clusters running smoothly.

The Special Challenge of Operations on Distributed Systems

Distributed systems are operationally complex. Managers often ask me: if distributed systems are so good, why isn't everything a distributed system? A big reason is the operational load of a distributed system. With a nondistributed system, all of your data and processing happens on a single computer, so it either works or doesn't. With distributed systems, you deal with tens, hundreds, or thousands of computers all at once. A problem is no longer confined to a single process or computer. Instead, there could be several layers of issues spread over several different computers and technologies at the same time.

© Jesse Anderson 2020
J. Anderson, *Data Teams*, https://doi.org/10.1007/978-1-4842-6228-3_5

The many different computers are just the beginning of potential problems. Once we add software, our issues increase.

The first level of software to worry about is the part that runs the distributed systems themselves. These distributed systems frameworks often require different processes and services to be running continuously. If these processes aren't running or functioning correctly, there is downtime. These processes run the gamut of finicky to resilient to "let me tell you about the time I spent 6 weeks trying to track down the problem." The operations team bears the brunt of this pain, although everybody involved with the service feels the external pressure during system failures.

The second level of software that operators need to think about is the code that the data science or data engineering team wrote. In production, you'll find out how good— or bad—that code is. Your operations team will have to support it.

The operations team may vary from one organization to another in their knowledge about the custom software the teams wrote. Remember that when there is a problem with your own software, you can't google for the answer or look on Stack Overflow. The knowledge is all internal and even tribal in nature. The more experience the operations team has with the internal software, the more they'll be able to support it. Without such knowledge, the most trivial operational issues will go directly to the person who wrote the code, whereas ideally, the operations team should have enough knowledge to at least attempt to troubleshoot the problem. This really put the onus on the organization to provide the operations team with the right amount of help and expertise.

Let's not forget the data aspect of an operations team. The operations team has to be familiar with the data being used. This includes information like the expected amount or size of the data itself, the type of data being sent, and the correct format of the data. Such information helps the team plan resource allocation.

In most small data organizations, the database serves as the repository for all data. This both centralizes data and makes it easier to keep track of its types and formats. With distributed systems, this becomes more difficult. Sometimes, there's no human-friendly GUI that helps the team look at the data and verify it. There isn't a central place or specific constraints put on the data to validate that it falls within particular tolerances or even that it is correctly formatted. These hurdles to dealing with data really change how the operations team works and thinks. With databases, operators were able to offload some operations and data checking onto the database itself, but now the operations and data engineering teams have to deal with it manually.

Some teams are forgotten when things run smoothly. This can happen to operations teams where things are perceived to be running so smoothly that the organization doesn't think it needs an operations team anymore. Operations may not be the most glamorous thing or even regarded as a cost center, but their contributions to data teams are absolutely necessary.

It's really foolhardy to undervalue what operations do. What happens to the organization's customers during downtime? Is there a massive repercussion if the data products are unavailable for hours or days? What happens if data were to be lost?

Some organizations think that running in the cloud—perhaps with managed services—negates the need for operations. Even with cloud and managed services, someone needs to make sure the software is running and operating correctly. Without consistent operational excellence, the data products may fall into disuse because they're operationally unreliable.

Job Titles for Operations Teams

The actual job titles on an operations team are more varied than other teams. Sometimes, this is due to what the organization itself has historically called the members of the operations team. Some organizations have increased the purview of the operations to include other tasks or raised the technical bar for the team.

The most common titles are *operations engineer* and *systems engineer*. The difficulty with these titles is that they don't connote the increased complexity that data and distributed systems place on the duties of an operations engineer. You will want to make it clear in the job description that their responsibilities will include data and distributed systems.

Another common title is *site reliability engineer,* or *SRE*. This is more common in organizations that are trying to follow more of the Google and LinkedIn model for operations teams. These teams are bringing more programming and software engineering practices to bear than your average operations team. Often, these SREs have been software engineers in a pre\vious team or role. These engineers focus on automating tasks that are done manually or ad hoc.

Finally, some organizations are more direct in their titles. They will actually use the name of the technology as part of the title or specify that someone does the operations for big data. These titles are *[some technology] operations engineer* or *big data operations engineer*, respectively. These people can be a great fit for the operations team, provided they have experience with the technologies they're expected to support.

What Skills Are Needed?

The skills needed on an operations team are

- Hardware

- Software/operating systems

- Distributed systems

- Troubleshooting

- Security

- Data structures and formats

- Scripting/programming

- Operationalization best practices

- Monitoring and instrumenting

- Disaster recovery

We'll discuss each here.

Hardware

Massive data can push hardware to its breaking point. This breaking point can range from things that are easy to fix (upgrading the CPU) to things that are difficult to fix (inadequate RAID controller firmware). An operations team running an on-premises cluster will have to handle everything on the hardware side.

In the beginning, this will include buying the hardware and maybe even racking and stacking. As the cluster gets up and running, operations engineers will be responsible for all of the hardware issues, such as hard drives and RAM. They're responsible for the troubleshooting of hardware issues.

Some people believe that running the cluster on the cloud makes all the issues go away. Cloud providers alleviate the need for directly interacting with the hardware, but merely using the cloud doesn't remove the need for troubleshooting hardware. Using the cloud actually increases the level and difficulty when you have to troubleshoot hardware. These troubleshooting expeditions are more opaque because the cloud providers don't list out all of the equipment that the virtual machine instance is running on or using indirectly. Other hardware troubleshooting could be a mix of debugging the hardware limits that the cloud provider imposes and the software rules in place for managed services. You'll find this out the hard way as you start to deal with issues like throttling.

Software/Operating Systems

Operations teams will be supporting primarily three types of software: the software your organization wrote, the software your programs need to run, and the operating system your applications run on. Supporting each one of these types is difficult, because big data problems and the ensuing hardware load will cause the software to break in unexpected ways.

Your organization's software will have bugs. It will be up to the operations team to figure out which bugs are due to your organization's software and which are caused by another vendor's software.

The software your organization needs to run is just as essential as anything else that needs to be supported. For many distributed systems, you need to use a Java virtual machine (JVM). The JVM itself can be an odd source of issues. For example, vendors will qualify a JVM's support of their technology down to the granularity of the build number. Other vendors haven't found a JVM that works 100 percent with their technology. The operations team will want to know about issues like these and the ramifications of not using the supported versions of JVMs.

Distributed Systems

I've mentioned how difficult distributed systems are. Now we'll look at distributed systems through the operational lens.

Distributed systems talk to each other. A *lot*. Sometimes this talking works, and other times it doesn't. Well-architected distributed systems will have mechanisms to retry failed communications. If these errors bubble all the way to the logging system, the operations team needs to take a look.

Since processing or software doesn't run on a single computer, tracking down the interplay of issues can be difficult. The computer reporting the problem may just notice the manifestation of the problem (in short, it's the messenger) rather than the actual source of the problem. Distributed systems lead to the creation of elaborate systems to track logging messages or entire products to monitor the life cycle of a distributed system. This need for proper logging is the nature of the distributed systems beast and is something the operations team should plan for.

Troubleshooting

There's a leap from the knowledge of something to the knowledge of how to troubleshoot it. Troubleshooting skills really kick in when you can't fix the problem by just rebooting computers or starting up fresh cloud instances. It takes a really systematic approach to eliminate potential sources of problems and identify the variable that really matters. It also takes a dogged determination to figure out and fix the problem.

The difficulty for operational troubleshooting is how many pieces or variables the problem has. These include

- Hardware

- Operating system

- Networking

- Distributed systems framework

- Your organization's software

- The software programs and libraries your organization's software relies on

- The data being sent or read

Your most difficult problems will be when the sources of the issues stretch across several pieces. This is when the troubleshooting skill is vital and put to the extreme test. In these situations, the operations team will find out whether it genuinely has troubleshooting skills.

Security

Security, usually, is primarily the purview of the operations team. Security in distributed systems can generally be broken down into encryption at rest, encryption over the line, authorization, and permissions.

The data engineering and data science teams often don't think about the security aspects of systems. They usually don't have an operational mindset and aren't thinking all the way through the system design to the security portion. It's up to the operations team to make sure that the other teams are adhering to security best practices in their code. For some distributed systems, the code itself configures the security and encryption of the access. The operations team may want to configure the distributed systems to prevent any insecure access.

There is a fine line between keeping things secure and making things so secure that no one can even access the data they need. These oversecure environments tend to create shadow IT where people start to start up their own, less secure hardware that bypasses security restrictions. By being too secure and paranoid, some organizations become less secure.

Data Structures and Formats

An operations team has to deal with data, the structure of the data, and the formats of files.

If an operations team doesn't understand or know the data they're putting into production, they will interrupt the other data teams with problems too often. An error might have been logged simply due to a bad piece of data or data that didn't fall into the correct range. This is really where the operations team has to make sure they hand the data engineering and data science teams useful metrics and sane error messages.

Operations teams often deal with binary data formats. The inability of a human to read binary formats is sometimes cited as a reason to favor string formats instead. This is not a valid reason. The data engineering should be creating utilities and tools to make binary data human-readable. Binary formats are actually better for operations teams because there is a more fixed and specified layout for the data.

Scripting/Programming

The need for programming skills is less certain for operations than for the data science or data engineering teams. Some operations teams don't need any kind of programming, while others could benefit from command-line scripting, and others need to program with dynamic or compiled languages. In general, the more programming the operations team has or understands, the better. The knowledge will help them cover more operational issues.

I've found that some operations engineers can read code, but can't write their own. For some issues, an operations engineer can just start reading the code to see whether the problem is a bug in that code. You'll want your operations engineers to have enough skill to at least sometimes be able to resolve issues so that they don't have to ask data engineers every time for help.

Operationalization Best Practices

Distributed systems have a plethora of configurations and best practices. It's up to the operations team to enforce best practices in the access other teams have to distributed systems. Distributed systems usually have sane or low-end defaults out of the box. When running on higher-end computers, these low-end defaults can prevent the distributed system from taking full advantage of the underlying horsepower of the computer. Other configuration changes can improve the performance for a specific use case. The operations team will need to understand both what the use case is and how to tweak the configuration to make the distributed system perform as fast as possible.

Monitoring and Instrumenting

The operations team is the one who has to figure out problems in production and keep things running. Other teams aren't responsible, and they usually don't think about what it takes to troubleshoot something in production. It's up to the operations team to ensure adequate monitoring and instrumentation of the programs in production to troubleshoot and recover quickly from problems. For some teams, the operations team can only call out the need for improvements and will need to work with the architects and developers for the actual implementation.

This could entail the operations team examining the code during code reviews. The operations team will look for two main issues that other teams probably aren't looking after.

- Error conditions and exception handling: Is every significant potential error logged? Is the piece of information or the state of the program logged to help the developers reproduce the issue? Does the error have a human-readable statement of where or how the error was thrown?

- Process instrumentation: Do you know how many messages or objects are sent or processed? How long did it take for a message or object to be processed? How long does it take for an external call like a database or RPC call to finish? Can you determine quickly whether the program is running correctly when it's in production?

You need to be able to answer these questions in order to gather enough information to re-create errors and fix them and to have the baselines to know when a process is misbehaving.

Disaster Recovery

Disaster recovery is required when an entire data center or cluster is inoperable. This inoperability could be due to a natural disaster, a power outage, a problem with the data center itself, and so on. These scenarios cause all or a significant part of the system to go down. In any case, companies need to have a plan to deal with a disruption that takes out more than a few computers.

You can't wait until a disaster happens to start copying data. The data needs to have already been copied to an entirely different place as part of disaster recovery planning. From the distributed systems point of view, each technology will have different abilities and built-in handling of replication.

Disaster recovery represents a whole extra level of difficulty and complexity for an operations team. Often, the designers of distributed systems or the internal architects haven't ever thought of how to deal with disasters. It may be up to the operations team to start that conversation. This conversation may include the distributed systems technologies themselves, as well as upgrades to the internal system architecture to handle disaster recovery. It will also cover the organization's code and business process. The code or a business process will have to dictate what happens in the event of a disaster.

When teams start using distributed systems, they often leave disaster recovery as a second release or even fail to address it in a proof of concept. Worst case, the technologies that were originally chosen may not support or deal well with disaster recovery because it wasn't a primary concern during the evaluation phase. This will be technical debt that the operations and data engineering teams will need to address.

Establishing Service-Level Agreements

The operations team is responsible for maintaining service-level agreements (SLAs). A service-level agreement defines the amount of downtime that is acceptable for a system to have. They can also specify how fast the response time must be. Many organizations break down their SLAs by the distributed system and the software running on that system.

The actual service levels will depend on the organization and how fast the data is processed. This processing or access could be batch-oriented, real-time, or specific to a service.

Batch SLAs

Batch SLAs are less stringent. In general, a batch SLA is measured in many hours to days. The system could be down for hours during the business day. This isn't ideal, but there usually aren't terrible ramifications because the data can be processed later after the operations team restores service. As part of digging yourself out of the resulting hole, many tasks and processes will need to catch up.

Batch SLAs can differ depending on the time of day, day of the week, or time of year. For example, the SLA could be focused on not having downtime during the business day. After the business day is done, the stringent SLA isn't necessary anymore.

Real-Time SLAs

Real-time SLAs are much more stringent than batch ones. Real-time systems often serve as the first point of entry for data. If the system is down, data processing can be delayed, or—even worse—important data can be lost. Companies with real-time systems need to make sure their real-time SLAs meet the business needs. Real-time SLAs are measured in minutes to low numbers of hours.

Moving from batch SLA to real-time SLA can be when an operations team gets into trouble. They're used to the less stringent SLAs of a batch system where there is a more elastic safety net. With real-time SLAs, the number of systems affected is much higher. In general, a real-time system feeds many different downstream systems, including both batch systems and other real-time systems. An outage of the real-time system causes a cascading failure to other systems.

The operational excellence needed for an SLA can help you figure out whether you actually need a real-time SLA. You need one if it hurts to have any downtime in the system. Otherwise, you may not have a real need for a real-time SLA.

Specific Service/Technology SLAs

Your SLAs may be different for specific services or technologies. For example, a database may be running a website. If the database goes down, the entire site could stop working. The operations team would identify the database's uptime as critical and establish an SLA accordingly.

Some services or technologies are outward-facing and may be critical partner services. If the service goes down, the method of moving data between the two companies could go down. In addition to websites and databases, there may be other important customer-facing services with corresponding strict SLAs.

Organization Code and Deployment SLAs

Your organization's code has its own SLA, apart from the services run by the operations team. You will need to make sure that the code your organization wrote has an appropriate SLA that applies when you update code that's running in production. The amount of time it takes to update the software could cause downtime. The team will need to figure out the best ways to upgrade the production system without causing downtime, or at least while minimizing downtime.

The underlying distributed systems software itself also requires updates. The operations team will need to establish a way to upgrade the software without incurring downtime. For critical systems, this means having the operational excellence to upgrade while keeping the system up. Some of the most challenging upgrade paths are when the operations team is years behind the current version of all software. At this point, the necessary upgrade will require both operating system upgrades and software upgrades. Because the software is so old, there probably isn't a direct path to upgrade from the old version to the latest version. Instead, the team has to install several major versions on the path to eventually get to the latest version.

Typical Problems Faced

Troubleshooting is usually the most demanding and high-pressure task for the operations team. It may not be the place they spend most of their time, but it will be critical to meeting business needs consistently.

This section discusses the main problems operations teams face, in the approximate order of importance. Each issue that an operations team faces may involve one, some, or a mix of all of these topics. You'll gradually have to peel back the layers to find what's really the issue. These groups of problems are what make distributed systems operations difficult.

Your Organization's Code

Your own code will be the primary source of your operational issues. If something isn't working, your code should be the first or second place you'll look. That doesn't mean that the operations team should pass off the operational issues directly to the data engineering team. Instead, the operations team should verify or have a good deal of certainty that the problem lies in the organization's own code.

There are excellent and consistent reasons to start here. For one, distributed systems code is complex, so errors are easy to insert. This is made worse when the people writing the code are new to the distributed systems arena. They may lack the experience to write the code correctly.

It's crucial for the code to have sufficient unit tests and integration tests before going into production. Without enough tests, it will be more difficult to figure out if the issue was fixed and didn't break something else in the process of fixing the bug. The operations team may need to really push the development teams for better testing to make their lives easier.

Troubleshooting issues are exacerbated by "throw over the fence" interaction between the operations team and the others. The operations team can feel as though they're receiving the brunt of untested code or code that is really buggy. These sorts of interactions lead to friction and distrust between the data teams.

Data and Data Quality

Some operational problems result from problems with the data itself. This is the first or second place you should look. Remember that data teams are creating a mix of software and data. A problem with data is just as dangerous as a problem with the software.

There could be several different problems with data. For instance, a single piece or an entire file of data can be corrupted. As a result, the portion of data or file may not deserialize properly. When troubleshooting the issue, you'll need to figure out whether the problem is from data and then figure out why the problem happened in the first place. By figuring out the source(s) of the problem, you can prevent it from happening again or at least document the scenario(s) leading to data issues.

There are other issues in which the data deserializes fine, but the data doesn't conform to what you were expecting. Perhaps some code wasn't written defensively enough, or the schema wasn't well constructed. For example, if the code casts a string to a number, producing a number outside of an acceptable range could give your program fits, but the problem may not be logged well. The operations team will want to work with the other data teams to know about implicit or explicit assumptions of fields.

The operations team may need to lobby hard for the creation of custom tools to help them check data issues. This could include scriptable tools that look at a part of a file or stream, deserialize it, and output the results. Most data engineering teams will not think to create these tools or believe it is important enough to spend time creating them. The management team will need to treat logging and tools as a critical feature and make sure that the data engineering team has enough time to add this to the code.

Framework Software

Distributed systems frameworks vary quite a bit in their production worthiness. If they are mature, it will have (hopefully) fewer bugs and better tooling to help you figure out issues. Some systems can be really troublesome. Other systems work really well, but 0.1 percent of problems will give you severe trouble or outright data loss. There really is a wide gamut.

Some problems in your own code will appear to be problems with the framework software. This is often due to the abuse or misunderstanding of what the framework is really used for or how a particular feature should be used or implemented. Sometimes these issues can be worked around with more hardware—usually an increase of RAM— while others require code fixes. Other problems can be inherent to the architecture of your distributed system, and no amount of workarounds will be able to fix the issue.

Hardware

Hardware can cause problems on rare occasions. The two most common hardware issues are rooted in buying cheap hardware and bugs in the firmware.

Real big data problems will push hardware in ways that the vendor may never have tested. Although the vendor's specifications may say the equipment can perform to a certain level, they may have never actually tested it to that load or tested that load for a long enough time. This leads to firmware and overprovisioning bugs that manifest only at scale.

Hardware issues are challenging to diagnose, primarily when they are related to networking. Some of the problems most difficult to figure out happen only sporadically. It can become virtually impossible to understand the exact situation that repeats the issue. This is why you really need a high-performing operations team.

Staffing the Operations Team

Now that we have an idea of what skills are needed on the operations team, we'll look at where to find the staff. Along the way, I'll point out some shortcuts that tempt managers, and what dangers lurk in these shortcuts.

The Need for Specialized Training for Big Data

Often management will try to put systems and network administrators onto data teams or simply change the title of existing individual contributors without actually giving the person the new skills needed for big data and distributed systems. This is a way to make your solution fail in production. If you're staffing your operations team with people who have previously only worked with small data systems, you need to give them the training to learn new skills related to distributed systems.

Going from the operation of small data systems to distributed systems is an increase in complexity. When you add in real-time distributed systems, the rise in complexity goes even higher. The operations team will need to understand each piece of the distributed system and how each piece communicates with the others. Without this understanding, operations teams won't be able to keep the operational issues from bubbling up to the data engineering teams.

Retraining Existing Staff

Creating a new operations team could include hiring new people from outside the company. More often, organizations look internally to make use of their existing operations staff.

Management should take care to provide people who are new to distributed systems with the resources to learn the new systems. It is a nontrivial task to learn how to install, configure, and troubleshoot these systems. A team that hasn't been provided the right resources will be unable to fix or prevent operational issues.

Operations

Your existing small data operations teams are the first places to look. Some of these individuals may have already shown an interest in learning how to operate distributed system frameworks.

These individuals will need to have an interest in data. Their current small data-focused may not have dealt with data or likely doesn't deal with data at the level needed by a data team's operations. An interest in understanding formats and other data issues will really help in diagnosing whether an issue is related to data or software.

DevOps Staff

Your DevOps teams task are another place to look for potential operations team members. In my interactions with DevOps teams, their skills skew more toward operations than software development. They may have better scripting abilities than most operations people. Using these staff may allow you to offload some of the automation and monitoring to the operations team instead of having the data engineering team do it.

But an operations team isn't just a DevOps team with a different title or slightly different purview. The two main differences center on data and distributed systems. DevOps teams often don't understand data at the level they should for an operations team focused on data. Distributed systems really are a specialization with specialized skills.

Just like the small data operations teams, the DevOps individuals will need to have an interest in data. And just like other operations roles, the DevOps teams will need extensive training in handling the operational issues of the distributed systems.

The development side of DevOps may be more difficult. If the DevOps team skews heavily toward the operations side, the team may never have learned the programming languages required for the code development. An operations-skewing DevOps team will have a great deal of difficulty with architecting and writing the necessary code to create data products, which is the usual root cause of operations people's difficulty with development on distributed systems.

Why a Data Engineer Isn't a Good Operations Engineer

I believe that operations are as much a mindset as a skill. This mindset of operations becomes apparent when you start to compare operations and software engineering. Remember that most data engineers come from software engineering backgrounds. They inherit all of the assumptions and training—some good and some bad—that come with being a software engineer. I wouldn't put most software engineers in a production system. Unfortunately, when a company wholly lacks an operations team, the data engineers are often the ones to fill in when it comes time to fix an issue.

I've seen software engineers and data engineering fumble with trying to do the simplest operational task, such as making a BIOS change or updating some software. This isn't to say no data engineer could ever be a good operations engineer. It's more a word of caution that data engineers may lack the mindset of troubleshooting production systems and end up like a bull in a china shop.

Putting a data engineer into an operations role may make them quit because they want to write code, not support running systems. Data engineers want to focus on creating new software rather than maintaining and monitoring it.

To make this more confusing, some data engineering teams are practicing DevOps. When a data engineering team practices DevOps, they completely take over the operations role and responsibilities.

Assigning a data engineering team to DevOps is very different from having a regular DevOps team take on operations for big data projects. Because the data engineering team is already familiar with the architecture and programming of distributed systems, they will have a much easier time with the operations of distributed systems. This doesn't negate the need for the data engineering team to be educated on how to operate the distributed systems being put in production; they will just have an easier time learning the operational rigors.

Management should exercise caution because doing DevOps doesn't magically make some of the really difficult parts of operations disappear. Instead, management should evaluate whether DevOps is the right fit for the team and organization.

Cloud vs. On-Premises Systems

Cloud usage really changes the work of the operations team. Some organizations are keeping their on-premises data center, while others are going for a hybrid approach that uses both their on-premises data center and the cloud.

If your data and virtual machine instances are using the cloud, most of the troubleshooting hardware needs to go away. Mind you, there are still hardware issues on the cloud you will need to troubleshoot, and these problems will be far more opaque than those of on-premises hardware. But the majority of troubleshooting will be focused on software.

For hybrid clouds, and when initially transitioning to the cloud, data movement will be an issue. The operations team will have to make sure the data flows securely and consistently. Depending on the technologies being used, this could be a manual process or an automatic process.

Managed Cloud Services and Operations

It's interesting to see organizations transition to the cloud and use it only as an outsourced infrastructure. In my opinion, the real value in the cloud comes from using managed services, both the ones offered by the cloud vendors and the ones hosted on the cloud vendors. Managed services allow organizations to use new technologies without having to teach their operations team how to manage and operate a brand-new technology. This allows the data engineering team to really focus on the best tools for the job instead of the ones that the organization can operate.

Organizations on the cloud should be offloading as much operational load as possible onto the cloud provider or managed solution. Some organizations choose not to use cloud providers because they perceive that they'll suffer from an increase in prices or from a vendor lock-in. A correctly used managed service should pay for itself with decreased operational overhead. Vendor lock-in can be an issue but can be reduced or mitigated with specific planning.

An opposite problem sometimes comes up with the transition to the cloud. Organizations using managed services tend to think they don't need an operations team anymore. They are wrong. Using managed services means your operations team can be smaller or lowers the bar for a team to do DevOps, but it doesn't mean you can eliminate the operations team completely. Your software will still fail, and you will even need to troubleshoot issues. You still need to have an operations team in place to figure these issues out.

CHAPTER 6

Specialized Staff

For the times they are a-changin'

—"The Times They Are A-Changin'" by Bob Dylan

As organizations become more mature in their use of data, they start to need more specialized skills. In this chapter, we look at two advanced areas of staffing: DataOps and machine learning engineers. Both skills draw on software engineering, analytics, and data science.

DataOps

A DataOps team is a cross-functional and mostly—if not entirely—focused on creating value through analytics. This cross-functionality allows the team to create end-to-end value from the data because they can take raw data and create value from it.

A DataOps team has data engineers, data scientists, data analysts, and operations on it. Having data engineers means the team can create data products efficiently. By having data scientists and data analysts, the team can create the full gamut of analytics that ranges from simple analytics all the way to full model creation. The operations component allows the team to put their product in production and have it supported or have the production access to use production data or cluster resources.

Just because the team can create data pipelines from the very beginning doesn't mean that they should always be creating things from scratch. They should be leveraging other data pipelines created by the data engineering team.

© Jesse Anderson 2020
J. Anderson, *Data Teams*, https://doi.org/10.1007/978-1-4842-6228-3_6

Continuing with the reusing strategy, they may not code every pipeline. DataOps teams may use more tools that allow pipeline creation without the need for code. Instead, these distributed systems or technologies will create analytics or data pipelines without the need to code them from scratch. The DataOps team would focus on speed of action rather than optimally engineered solutions.

This leads us to talk about the critical distinctions between DataOps and data engineering—the speed of cycles. Data engineering—like most software engineering—works on more prolonged periods. This is the outward manifestation of the sheer complexity and time-consuming nature of distributed systems. More data engineering team velocity improves these timings, but that's usually tapered down by the data engineering team, creating more and more difficult data pipelines. DataOps realizes that business conditions change quickly and each business is competing with others to discover and respond to change; organizations with well-established data teams call for new models to be production-ready and deployed as fast as possible. To accomplish this, DataOps works on a much shorter cycle. The DataOps team can take the data engineering team's data products, work with the business to create the analytics, and create the specific business value from the data products. This shorter cycle is really what the business side needs to act on their data.

The Trade-Offs Made in DataOps

Some projects have to get done in the timeline of about a day—perhaps even in hours. It's happened that a VP or CxO asks for an analytic for their meeting the next day. A DevOps team can turn that workaround. Asking the data engineering or data science teams to do it forces them to completely stop their work, change contexts, and start hacking. This consistently creates delays and issues for the team—and they probably lack the skills and mindset to meet the deadline, because their whole careers have been devoted to careful, bulletproof coding.

The key principle of DataOps is to focus on speed of development and release, rather than on optimally engineered solutions. DataOps staff don't try to substitute their skills for other teams or take on everything done by the other teams. Instead, the DataOps staff may let the data engineering staff create new pipelines and then apply DataOps skills to allow fast updates and releases. For some data sources, there may not be a preexisting data pipeline, and the DataOps team may need to create it themselves. DataOps may use tools to create pipelines without having to write new code.

Thus, data engineering does its job first, at a pace comparable to other software engineering teams. They need a lot of time because distributed systems are complex and hard to use correctly. DataOps, on the other hand, works on a much shorter cycle. DataOps can take the data engineering team's data products, work with the business to create the analytics, and create the specific business value from the data products. If they're having to trail blaze, they may create a data product that the data engineering team comes back and makes more production-ready.

PROBLEMS WITH GARTNER'S DEFINITION OF DATAOPS

Gartner, which is frequently quoted among business leaders in technology, offers its own definition of DataOps:

> Data ops is the hub for collecting and distributing data, with a mandate eto provide controlled access to systems of record for customer and marketing performance data, while protecting privacy, usage restrictions and data integrity.[1]

In my reading, Gartner's definition does not adequately distinguish the new skills brought by DevOps from the traditional role of a DBA of the data warehouse team. The definition actually suggests a resistant approach: keeping users away from data.

My critique is important because, historically, data teams have used their responsibility for privacy and for safeguarding data to hold back valuable uses. A turf war between data users and data warehouse teams has often broken out over this conservativism. What we want is a definition that emphasizes agility and increased access to data.

I mention the Gartner definition here only because you're likely to see it used elsewhere, and I want to make sure you have a more positive definition to work with.

---

[1] www.gartner.com/en/information-technology/glossary/data-ops

[2] Data scientists often come from highly academic backgrounds. In the academic world, ideas and models have to stand up to in-depth scrutiny like a Ph.D. dissertation defense. Conversely, the

THERE ARE THREE MORE DEFINITIONS OF DATAOPS THAT ARE OUT THERE

One is that DataOps is just DevOps for analytics. In this definition, we're taking the DevOps movement and saying we need to expand it slightly to include data. This thinking says if we just release faster, we can increase the quality of our data pipeline. DevOps focuses on value generation via software engineering and improved release cycles. A DataOps team should be focusing on value generation from data. The DataOps team can only create value from data when the data is of enough quality to make it usable. Ideally, the DataOps team is as self-sufficient as possible to self-serve data products and productionize any outputs.

The second definition is a practice for the data engineering team. This definition takes your data engineering team and says the need to apply DevOps principles. This means that your data engineering team needs to learn or add people with operations and value creation from data—usually analytics. Combining these two new properties to a data engineering team is a nontrivial task. On the value creation side, the data engineering team would need to learn how to analyze the data and create usable insights from the data. Once again, having taught data engineers extensively, this isn't a common trait. If the value created from data is simple or straightforward math, a data engineer could do it. I'd argue that if your analytic products are that simple, you should really be shooting higher and may actually be wasting time and money.

The third definition is a superset of DevOps, Lean, Agile, and data management. This definition allows the DataOps team to optimize the beginning to end creation of data products. The team would have control over the provisioning, data quality, and operations of data products.

Finding Staff for DataOps

In contrast to other teams and titles, I don't think there will be a consistent DataOps Engineer title. Because the DataOps team is cross-functional, you need to distinguish the data engineer, data scientist, data analyst, and operations person. The people you hire or draw from other teams will probably, therefore, retain their original titles.

Organizations that already have the data science, data engineering, and operations teams described in the previous chapters tend to find DataOps staff in one of the following ways:

- From the data engineering team, by teaching staff the necessary software engineering skills

- From the operations team, by teaching people the necessary data and analytics skills

The real question is who has experience in software engineering. It's important to be mature as a software engineer, and picking up the other necessary skills after that is not so hard.

The amount of change for data scientists and data analysts can vary from minimal changes to a much larger group of changes. Data scientists and data analysts are often used to working on shorter time frames. If they aren't, they will need coaching or training on which corners to cut to get something out fast. If the data engineers are using no code or low code products to quickly create data pipelines, the data scientists and data analysts may need training on how to use it effectively.

The Value of DataOps

Most organizations get started with data because they've heard of the near-miraculous things data science has achieved and assume that data scientists are all they need. The managers probably don't understand much about the requirements of dealing with data, so they expect their data scientists to do everything from finding the data to creating business value from data. As we've already seen in this book, effective data analytics is a team effort requiring many skills.

Some analytics and data engineering organizations are propped up by the sole efforts of one or two people. This individual heroism represents all sorts of risks to an organization. From a technical perspective, this heroism involves more duct tape and uncertain hope than you may like. For a short-term analytics project, that's fine. For long-term or enterprise-grade analytics, this approach creates copious technical debt. This sort of heroics doesn't scale, and the organization will find this out the hard way once that person leaves. Constant individual heroism is often a sign of deeper technical and personnel issues in the organization.

DataOps should give the organization a way to deal with the issues—and issues will come up—while taking unreasonable burdens off of individuals' shoulders. The DataOps team should be focused on gradual and continued improvement until there isn't a need to burn the midnight oil anymore.

Relationship Between the DataOps and Data Engineering Teams

At larger organizations, the DataOps team is separate from the data engineering team. This division of labor requires them to work out which team does which tasks, and how they will work together.

The teams tend to have different mentalities: the data engineering teams come from software engineering, and DataOps comes from a focus on speed to create analytics.

Because of the different mentalities, the DataOps team may think that the data engineering team is too slow and is over-engineering things. The data engineering team, in turn, can think that DataOps is playing too fast and loose with engineering rules and best practices. The data engineers may fear that they'll end up owning a poorly implemented, poorly documented, and failing system in production.

This is a fundamental tension you will have to work around. An engineer, especially in a large enterprise, thinks in a ten-year time frame. They feel that their job is to ensure that the program or system they develop can be maintained and work robustly for the next ten years. Operationalizing any new technology has an inherent risk that the engineers are trying to weigh. They're thinking of these issues because they been bitten by them before.

An analyst or data scientist on a DataOps team isn't thinking on these timelines of ten or more years. They're on a short-term schedule, and their only concern is whether the analytic works or to experiment with a new idea for the short amount of time it's needed—and this could be just minutes or days.

Thinking that everything is either a short-term or long-term problem is where DataOps and data engineering teams will get into difficulties. They must learn to balance speed with enterprise considerations. Sometimes, a short-term problem or solution will stay a short-term solution. But what happens when that short-term solution takes on a life of its own and becomes a long-term solution?

Furthermore, sometimes you will have to replace a short-term solution with a different, more long-term solution. Failure to recognize the need for a shift in time frame could lead to solving the same problem over and over, inefficiently, with short-term solutions.

Thus, the DataOps and data engineering teams should have clear communication processes in place to anticipate and handle the shift between fast, short-term solutions and more architected, long-term ones. Figure 6-1 suggests the conversations the two teams could have as they recognize the need for a shift.

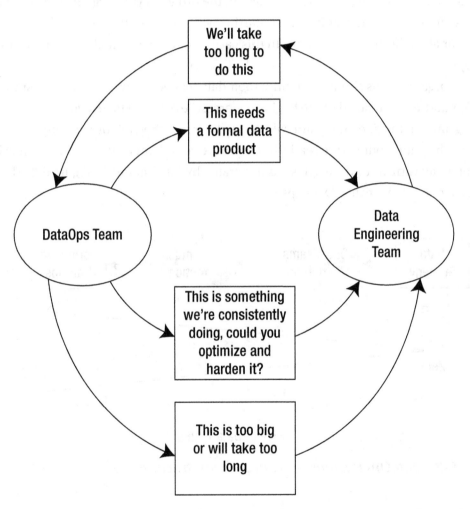

Figure 6-1. *A model for how DataOps and data engineering can work together*

There are still projects that need a conventional engineering timeline measured in months. As the complexity of analytics grows and the need for data products increases, you will need both your regular data engineers and your DataOps team.

When to Start Doing DataOps

I've found that attempts to institute DataOps too early in the organization's history will risk failure because the business is making too many organizational changes at once. The team fails because it's confused about whom to report to or what to do. Added to the sheer technical difficulty of the move to big data, the team may just cease to be productive.

You can consider starting DataOps early in the project if your organization already has cross-functional teams or if you are starting with a large data team and have enough people for all of the necessary teams. In this case, some staff can make lateral moves into DataOps.

Most organizations should wait until their data teams mature and hit the issues that DataOps addresses. Instead of the beginning pressure of trying to just get something working and out the door, the team will start to hear that they're not releasing tools fast enough. The data products are ready, but there are complaints about the speed at which the teams can turn around a request, as illustrated by the timeline in Figure 6-2. This is an excellent time to create a DataOps team.

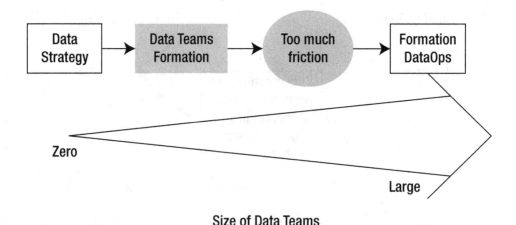

Figure 6-2. DataOps is a more advanced team structure

Machine Learning Engineers

Machine learning engineers handle a skills gap between data science and data engineer shown in Figure 6-3. The data scientist is strong in statistics and AI, whereas data engineering is strong in distributed systems and building pipelines. Each has some knowledge of programming and tools for using big data, but as the middle of Figure 6-3 shows, their skill levels drop way down. There's a gap that's difficult to bridge. The severity of this gap varies depending on the organization, the difficulty of the tasks, and the data scientist's own programming skills.

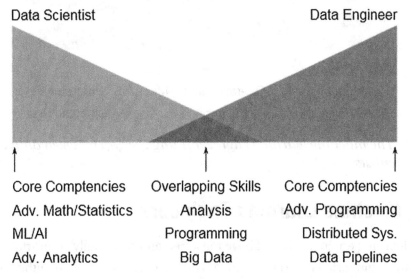

Figure 6-3. *A visualization of the gap between data engineers and data scientists*

Often someone needs to rewrite the data scientist's code before it goes into production. The change can range from a brief code review to a complete refactoring and rewrite. But it's not enough just to hand the code to an experienced programmer, because it embodies advanced statistics and a sophisticated machine learning algorithm. The programmer needs some familiarity with these data science skills. During this rewrite, the model may need just a tweak, whereas other models need to be rewritten completely. The job could include making the deployment of the model more straightforward and more automatic. Sometimes, there are concerns that are a mix of operations and data science where a new model needs to be verified as performing the same or better than the previous model.

This is the gap filled by machine learning engineers. They sit between the data engineering and data science worlds. They'll need to able to see whether a model is good enough for a use case vs. defending a Ph.D. dissertation.[2] At the same time, they'll need to know when and how to apply engineering rigor to a model. Figure 6-4 shows where the machine learning engineer fits into the work required.

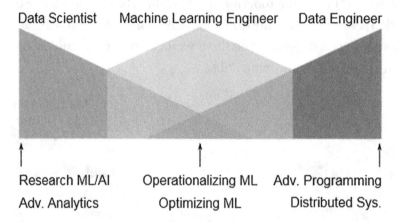

Figure 6-4. *The machine learning engineer filling the gap between data engineers and data scientists*

Finding Machine Learning Engineers

A machine learning engineer should have mastered and be equally comfortable with both software engineering and data science. Thus, they can come from either side of Figure 6-3: a data scientist who has experience with software engineering or a software engineer with experience in data science.

Data scientists who take on the job of machine learning engineer have often done their undergraduate work in computer science or a related engineering field. Some have worked as software engineers previously. In either case, they have a software engineering pedigree that most data scientists lack.

---

business world has a different level of scrutiny that is more of a "does it work for the use case" level.

The other background is a data engineer with an interest in data science. They've often shown a great interest in data science, and their schooling may have included a minor in math or some sort of math focus. Others lack a paper pedigree entirely but plan to advance through data engineering into a machine learning engineer position.

Where to Locate Machine Learning Engineers

If you have only a few machine learning engineers, you can place them in either the data science or data engineering team. I've seen machine learning engineers mostly in data science teams because the rewrites and improvements to the models are the most time-consuming parts of their job.

With a larger organization or a larger number of machine learning engineers, you may need to create an entire team of machine learning engineers. The machine learning engineering team will maintain a high-bandwidth connection and have extensive interactions with the data engineering team.

PART III

Working Together and Managing Data Teams

Part 2 covered who to hire for the three data teams you need. Part 3 will show how to coordinate data teams with the business to create something useful with the data.

Some people on the business and executive side like to think their job is to hire the people. Others believe that their job is done once the data strategy is completed. The reality is that successful teams are consistently and continually coordinating with their business counterparts.

Big data projects have many steps on the road to production. It's important to both know and follow the steps to be successful.

CHAPTER 7

Working as a Data Team

There is no life I know
To compare with pure imagination

—"Pure Imagination" by Gene Wilder

Standing between books and business needs comes the reality of implementing and creating data teams. In the real world, you have to handle egos, corporate politics, and the expectations staff may have in their first jobs outside of academic careers. This chapter looks at some of the challenges you have to face when building or expanding your big data efforts:

- Roles and relationships on teams
- Conveying corporate expectations while managing the expectations and egos of staff
- Avoiding unproductive distractions
- Fixing technical debt

Getting People to Work Together

Although companies often appreciate data scientists, who are the rock stars of big data, management tends to underestimate the importance of data engineering, which was defined in Chapter 4. Big data practitioners frequently say that getting data into a usable form—such as cleansing, data product creation, and the infrastructure to run the model on—takes up 80 to 90 percent of the job, and most of that work falls on the data engineer. But some organizations fail to account for data engineering at all or assume the data scientist can do it.

Part 2 helps you understand the actual roles each staff person should play. Some advanced roles—the DataOps staff and machine learning engineers discussed in Chapter 6—cross over and fill gaps between roles.

© Jesse Anderson 2020
J. Anderson, *Data Teams*, https://doi.org/10.1007/978-1-4842-6228-3_7

Staffing Ratios

When data engineers are hired, the ratio of data engineers to data scientists is often one to one. In reality, you'll need two to five data engineers for each data scientist. This ratio is so high because the 80 to 90 percent of the work for tasks fits into data engineering tasks rather than data science tasks. The ratio depends on:

- Level of complexity required on the data engineering side

- Data scientist's knowledge of data engineering skills

- Complexity and number of models

- Complexity and amount of data

The ideal ratio of operations staff to data scientists or data engineers can vary even more between organizations. The number of operations personnel you'll need depends on SLAs, your degree of operational efficiency, and managed service usage. It may be worth keeping a fairly constant ratio between the number of operations personnel and the number of data pipelines and models in production.

Getting your ratios right and making your teams productive does take some finesse. If in doubt, err on the side of hiring more data engineers—you'll find work for them, I guarantee!

Should Data Teams Be Separate or Together?

Smaller organizations can benefit by putting their data scientists, data engineers, data analysts, business analysts, and operations together in a single team. It makes no sense to break them up and put a manager over one or two people. But larger organizations will need to create separate teams—having a 50-person team doesn't make sense either.

Some managers hope that combining everybody in one team will prevent siloing, especially in highly political organizations. Unfortunately, the politics and ego battles may just resurface in different ways. Before assigning staff to teams, look at the critical issues behind your organization's political dysfunction.

It may work to keep data scientists on one team while combining data engineers and operations on another team.

Political troubles are particularly likely if data scientists, data engineering, and operations are in separate parts of the organization under different managers. See how high in the organization chart you have to go before there is a common leader to all of the data teams. At smaller organizations, this can be at the C-level. For larger organizations, this can be at the director-level, VP-level, or even stay at the C-level.

Try to structure the company so that the manager held in common is not high up. When there's a problem, you have to go all of the way up to get a decision. Similarly, any other issues, prioritization or capacity, or competition in vision has to go all the way up before a decision can be made. In highly contentious situations, this can be a constant and burdensome problem that really cuts down on productivity.

To be clear, I am recommending that data teams have a common leader that is relatively near them in the organization chart. This common leader can establish clear, consistent, and complementary goals across the teams.

But if teams end up in different organizations, it's critical for their managers to reach across the organization and create a rapport with the management of their fellow data teams. This task is the most challenging part of management.

The DataOps team, discussed in Chapter 6, is a special case because it's cross-functional by definition. Of course, even though it contains people from many disciplines, it still has to interact effectively with the other teams.

PRODUCT TEAMS VS. FEATURE TEAMS

Another way to characterize the organization of data people into one team or spread into different teams is the idea of feature teams (all together) and product teams, where product means anything that will be used by paying customers or key internal stakeholders. Product teams are inherently cross-functional, having a mix of product owners, data scientists, data engineers, front-end application engineers, back-end application engineers, and so on.

Product teams are popular because they encourage a focus on the delivery of the products essential to business outcomes. They avoid some of the potential issues with feature teams, such as building overly large and complex libraries that perhaps cover *every* team's needs, but are also bloated for any *one* team's needs. Put another way, feature teams can lose sight of organizational goals, the "big picture."

The drawback of product teams is the difficulty of specialists, like data scientists and engineers, to build organization-wide standards, best practices, and libraries. Hence, a tension exists between these two organization modes, product vs. feature teams.

What works best will depend on your organization. For large, mature organizations, having feature teams that standardize their specialties might be best, especially if new product releases tend to be assembling or fine-tuning existing components. When the whole invention is required, and standardization is not yet useful, product teams are best.

—Dean Wampler, Author and Head of Developer Relations, AnyScale

High-Bandwidth Connections

How do you overcome the hurdles of locating data teams in entirely different parts of the organization? How do you make sure they actually accomplish their work? The solution is to create high-bandwidth connections between the teams.

A high-bandwidth connection means that the members of the data teams know each other. They build camaraderie and respect by seeing that each other is competent in their respective fields. When one person contacts another, there is an assumption that the caller spent the time and effort to solve the problem themselves or search for the answer before interrupting another person.

This high-bandwidth connection can be created by embedding members of one team in another. In the organizational chart, the embedded member can have a dotted-line connection either to their old team or to the team that they're working on.

For example, a member of the operations team could be embedded permanently or temporarily within the data engineering team. This could be done to facilitate a specific need on the team or to fill a skill hole on the team.

Be careful when embedding someone permanently or long term on a different team. People may not want to dwell outside their home team for extended periods. It's useful to look or ask for volunteers before assigning them to a team. By asking for volunteers, you may find someone who already has an interest in cross-training or switching over to that role. Take care to address the hole left by the person in their original team. These are often senior roles and leave a large void that the previous manager will have to fill in personnel and workload.

An Iterative Process

Getting teams to work together is a matter of gradual repetition and improvement. Just like some agile processes, this succeeds by creating smaller cycles during which the team gets better. A data team can't fix everything all at once.

Teams that focus on getting everything right at once will be stymied in gridlock because they can't focus on an achievable goal. Instead, the data team should focus on getting a little bit better over shorter cycles. This effort will start to compound into being much better overall.

Politics, Expectations, and Egos

Too many projects languish in the political fights of the organization. It is crucial to solving this. Although I can't solve an entire organization's political problems in a book, I can offer some of the common problems here along with possible solutions. I often advise teams to solve the political problems that are just within their purview and groups rather than trying to fix broader issues.

Data Project Velocity

You may have heard about a concept in Agile project management frameworks called *velocity*. Velocity is the speed at which the team gets things done on the project. A sign of a healthy team is that their velocity will gradually increase as the team matures in their usage of the project management framework.

There is a similar velocity for data teams and a similar ramp-up time for distributed systems and their usage. The data teams initially need extra time, space, and resources to start getting things done with a distributed system. A sign of healthy data teams is that their velocity gradually increases. This means that the teams are getting more comfortable with the distributed systems. They're gaining the experience that allows them to create better code faster. Given enough time and effort, team members will eventually become project veterans.

Too often, project plans and management expect new data teams or new members of data teams to be experts on their first day, either when they are hired or after they been given some kind of training. Expecting expert output from a new team or member puts them into the untenable position of being behind from the beginning and gradually getting further and further behind, as they are unable to complete what they supposed to in the sprint.

The inability to keep up is where the team's velocity comes into play. If the team never gets the chance to catch up or adjust their project plan, they will fail.

On the other hand, if the team—despite all the time, resources, and help given—never increases their velocity, there is some problem affecting the team that management needs to address. The data teams' velocity is bidirectional to provide the individual contributors the resources they need and gives management the metric to hold the data teams accountable.

When calculating the velocity for a brand-new data team or for a member of a data team who is just learning distributed systems, you should expect it to take 3 to 12 months for the person to feel completely comfortable with distributed systems. This speaks to the sheer complexity of distributed systems and the copious number of technologies that data engineers have to learn. Each one of these technologies can be significant to learn in terms of time and cognitive effort.

Create a Crawl, Walk, Run Plan

Data teams can't go from 0 (beginner distributed systems) to 100 (expert distributed systems) overnight. The team has to get some velocity first.

Sometimes, I'll find a team whose manager or lead engineer went to a conference and watched an organization's presentation on their architecture. The person will then base their entire architecture, vision, and product road map on that presentation. What the person doesn't realize is that it usually took the organization presenting the results years to actually implement their architecture—they just don't say that during the presentation.

Chasing after another organization's implementation will get new data teams into trouble. They don't realize it's going to take years to implement the architecture they're trying to do. This makes the team unable to show any value for years.

Another common scenario is that the CxO—usually the CEO—has a grand vision for "success" with big data. The reality is that the organization isn't ready and can't achieve the CEO's goal. This forces the data team to find a way to create something that is somewhat doable. Projects that show no value for long periods tend to get canceled or gutted. For this reason, I highly recommend teams break down their use cases and backlog further. I call this further breakdown "crawl, walk, run." It lets a data team gradually get to 100 percent of the target while not taking forever to achieve any result.

In crawl, walk, run, the data teams stage out their backlog based on when the team's velocity is sufficient to accomplish the task. It's good for everyone to realize what is out of reach at each point in time. Each phase should set up the team to achieve the next step.

Spreading the Love

Most—if not all—credit for data projects goes to the data scientists. This is because the data scientists are often the public face of the data products. The data scientists are the ones working with the business to create the data product. In the perception of the business side, the data scientist did it all. If you look at the majority of organizations, the impression is that the data scientist accomplishes their task entirely by themselves and without help from others.

Thus, management must recognize other teams' work too. Data engineers and operators frequently complain that praise rains down only for the data scientist. They're particularly angry when they're understaffed (see the "Staffing Ratios" section earlier in this chapter), underfunded, and constantly knocking themselves out to meet the needs of the data scientists. The data engineers and operations team members may well quit and go to another company that recognizes their value. It's up to the team managers to educate upper management on the cross-functional nature of data teams and how others' contributions enabled the data scientist to accomplish their job.

Communicating the Value of Data Engineering, Data Science, and Operations

It is incumbent on management to communicate the value and importance of each of the data teams for your organization. Not everyone in the organization will understand the value created by data teams. This is where management needs to educate internally.

If a data team is well run, it may not make the waves that a poorly run team will make. The poorly run team will always be carrying out heroic saves and making noise because they're fixing a problem that they or others created. To show the value of a well-run team, that teams need to make their own noise by telling others that it's a good thing that they're not working all hours to fix mistakes.

Data engineering and—especially—operations won't get the credit they deserve unless management makes a concerted effort to educate others. Failing to garner praise will make other teams think that a person or an entire team isn't necessary to the success of the project. The reality is that key people often take their own competence for granted and don't know how to call attention to their accomplishments.

Data Scientists Need the Contributions of Other Teams

The consistent message of this book is that big data is a team effort requiring many disciplines, but data scientists sometimes don't get this message. Here are the barriers you may run into when trying to give them help from other members of the team.

If data scientists come from academia, they may never have to face engineering issues such as accessible and searchable data storage, maintainable coding, and rapid releases. In academia, the data scientist's value came from writing a paper, not creating a model that goes into production. The problem is exacerbated if management, dazzled by the promise of data science, also fails to understand the importance and difficulties of those tasks.

Being told repeatedly how awesome and unique a data scientist is can go to their heads. Couple this with freshly coming out of an academic setting, and you have a recipe for trouble. It calls for growth in the individual as well as a balanced recognition throughout the organization of all staff's contributions.

Data scientists are usually novice programmers. It may not be a good idea to put a data scientist's code into production. I'm not just picking on data scientists—this applies to anyone who is just starting out as a software engineer. Putting a new person's code into production can be a recipe for disaster. Keep in mind that data scientists are often self-taught coders, or base their knowledge on university classes that taught programming syntax but not software engineering. This provides a crucial realization of why their code needs to be rewritten or—at a minimum—reviewed.

Data scientists will chafe at having engineering processes being forced upon them. It's worth pointing out to them the difference between programming and software engineering. Programming is knowing the code or syntax enough to solve the problem with code. Software engineering is creating a solution that adheres to known software engineering best practices such as unit testing, source control, and correct usage of underlying technologies. To a newly minted data scientist, all of this engineering seems like wasted time and over-engineering. But there are some engineering processes that data scientists should be following. It's up to management to wade through ego needs to show the necessity of working together.

There is an ego component to telling someone their code needs to be rewritten. The data engineers, more experienced data scientist, or machine learning engineers doing the code rewrite will want to keep in mind that this can be a teaching opportunity for the data scientist. The data scientist isn't writing poor-quality code on purpose. Most often, they've never been coached about best practices or reasons why something is done a particular way in software engineering.

The person rewriting the code faces a difficult challenge. They probably won't understand the math or statistics being used in a data scientist's code. This is one place where a high-bandwidth connection between data engineering and data science is crucial.

Data Hoarding

Politics isn't just limited to people; it can extend to the data too. Often, companies refer to their data as being siloed. Sometimes, this siloing is the direct manifestation of politics, when one team hopes to gain an advantage by not allowing other groups to access to their data. When starting a data team, your first fight will be to wade through the siloing and politics to get data shared as needed around the organization.

Part of the solution to politics involves getting people to help you achieve your goals. Sometimes, it isn't enough to get pointed to the data. In these cases, you'll actually need someone's help to understand the data or what the data means. That could mean convincing that person's manager to assign someone to this task or just showing patience with a surly person who doesn't want to help you. Either way, you'll need to know that some vital data isn't held or stored in a pristine manner, and your data teams will need to rectify this.

Death by a Thousand Cuts

Getting something simple into production can be difficult. The difficulty doesn't just lie in technical barriers such as the complexity of the distributed systems but also lies in the ecosystem of the organization. It also lies in dealing with all the pieces it takes to get distributed systems running correctly. Each of these difficulties can lead to the death of the team's productivity. I call it a death by a thousand cuts.

The Curse of Complexity

At a big enough enterprise, you might struggle with the complexity of just setting up the project. This isn't limited to the infrastructure that runs the distributed systems. It also stems from simple tasks such as source control and build systems. Each enterprise will have slightly different nuances and quirks that are inherent to the organization. These quirks aren't searchable or aired on Stack Overflow. They're usually tribal knowledge, and fixing them requires finding a senior person who has hit the same issues before.

Most data pipelines use many different technologies, complicating both pipeline creation and troubleshooting. When something goes wrong, you aren't just trying to figure out a problem with a single technology, as usually is the case in small data. Instead, you have to figure out which of five complex technologies could be the problem. If you're really unlucky, you'll have a problem with several of them at once.

Thus, troubleshooting in a data team's technology stack takes much longer than in other parts of a software project. It can be a massive time sink to go in and debug a system. This can quickly lead to the death of the team's productivity, as they spend their time trying to figure out what went wrong and where.

Going Off the Beaten Path

Going off the beaten path means using technology in a way it wasn't originally designed for. This is a time when the team thinks it's possible to use (abuse) a technology a certain way but lacks the ability to point at another organization using it the same way. Going off the beaten path happens for several reasons. The team's use case could require a new or less well-known technology to implement it. Some teams just want to use the latest and greatest technologies because they're cool and hip. Immaterial of the root cause, the manifestation is still the same.

When going off the beaten path, death by a thousand cuts can happen gradually. You can realize one day that the team has spent months fighting the same issue and never getting anywhere. The teams usually don't fix themselves without external help from an actual expert on the technology or use case. The knowledge isn't searchable on the Internet or found on Stack Overflow. It is a tribal knowledge that is centralized in a small group of people who have been using the technology for a more extended period or are the creators of the project.

Organizations facing higher technical risk will want to take specific actions to mitigate their risk. This should include help and support on the newer technologies. This support isn't limited to operational support—it should consist of support for the architecture and programming side too. Ideally, this should include training for the operations and data engineering teams on the technology itself.

Technical Debt

The creation of technical debt on data teams—introduced in Chapter 3—needs to be called out and known. When implementing a project using the common strategy of "crawl, walk, run," the team will deliberately create technical debt that will be gradually paid down in the next phase. The open recognition and acknowledgment of technical debt will help prevent self-flagellation or condemnation from others when the team can't accomplish a use case to the 100 percent level. For the business that is at a 0 percent level, a 50 percent level may be enough to get by for now. The remaining 50 percent stays as technical debt that the team that will pay down more in the next phases or iterations.

Management should be acutely aware of the technical debt that is being created. Managers will need to make sure the team is paying down their technical debt through concerted effort, whether created deliberately or mistakenly. If management actively or mistakenly prevents the team from going back and fixing the technical debt, the data team's confidence in management will diminish. Members of the team will perceive management as being irresponsible about the future and focusing obsessively on new features. Not fixing technical debt will eventually lead to massive rewrites of code that take a long time because the team was never given the time to pay down the technical debt iteratively.

CHAPTER 8

How the Business Interacts with Data Teams

Without you, there's no change

—"Without You" by Mötley Crüe

Gaining the maximum amount of value from your data requires constant interaction between the data teams described in this book and your organization's business side. A data product in and of itself doesn't create value. A data scientist can analyze data all year, but it won't make an impact until some business value is realized.

How does a business actually benefit from big data? It all starts with changing a can't to a can. Each of the data teams supports the creation of the value chain in their own way. For the business, the end benefit of the value chain could come from many things, depending on the organization and its forms of income. Sometimes, the business side knows what it wants from the start and sponsors a project or the entire creation of the data teams. Other times, the business involves bringing in sponsorship or partnerships in some fashion. In any case, the business is trying to connect a customer or client problem with a technical solution.

Business interaction is critical to ensure that its data products answer these questions:

- Does the data product as it stands actually meet the business needs?

- Did the data product create some value?

- Is the data product creation or completion within an amount of time that's acceptable to the business?

© Jesse Anderson 2020
J. Anderson, *Data Teams*, https://doi.org/10.1007/978-1-4842-6228-3_8

This chapter examines these interactions in the following areas:

- Sources driving change

- Where to locate the data team and its members

- How the business should interact with data teams

- Funding and resources

- Topics for interaction

- Rewards and measuring success

How Change Can Be Accomplished

Becoming a data-focused company will take a great deal of change. It is up to management to be at the forefront of pushing this change. How the organization makes the change will depend on which level of the organization is pushing for the changes: top down, middle, or bottom up.

Pushing from the Top Down

A top-down push means the change is coming from the C-level or another executive in the organization. Of the three ways of accomplishing the move, this is the least painful way of effecting the change.

The difficulty for executives will be to get alignment from the entire organization on the goals and strategies for data. Just creating a data strategy isn't enough. The executives will need to continue to push to become a data-focused organization. They can lead by example by pushing for relevant data when making a decision.

Top-down pressure can be most effective in ensuring that silos are being broken down. The executives will need to make sure the other teams in the organization understand the importance of working with the data teams.

The executives will also need to make sure that the proper resources are given to the teams. This includes creating a headcount for the new data teams. Other resources include any outside help that the team needs to learn distributed systems or specific consulting on the architecture. This prevents teams from getting stuck and not making any progress on the data strategy.

Middle Pressure Up and Down

Sometimes middle management sees the need for organizational change first. Their difficulty is that they're in the middle of everyone. They'll have to effect change by pushing both up and down.

I often find that architects are placed in the middle position too. They're keeping tabs on the technical and business needs of the organization. They may be some of the first people to see the can'ts.

This middle position is a difficult, but not impossible, place to effect change. The middle management will need to spend time convincing upper management that there is a need and business value to creating data teams. Middle management will also need to convince the individual contributors that it's worth spending the time and effort to learn the distributed systems. I highly recommend that the managers start by looking for the business value of creating data teams and dedicate most of their energy to convincing upper management of the benefits.

It may not be possible to convince everyone about the benefits, especially upper management. This can be a time where I recommend middle management lead with success on a project, for instance, creating a skunkworks project that solves one of the organization's can'ts. This project should clearly demonstrate the potential value and benefit if the organization were to just start investing directly in a data team. Most people—especially upper management—like to get behind a team that's already winning and showing promise rather than taking a chance on a project and team that's unproven.

Bottom Up

Although this book is written for management, I suspect that individual contributors will read it. I'm assuming that you're seeing the need for a data team, and you're trying to figure out how to get it done. This could be because management—both middle and upper—aren't interested or don't really understand the potential value of data. It will be challenging to push from the bottom up, but there have been people who made it work.

Individual technical contributors often lack the right wording or positioning to get things moving. Are you saying that we need to start using distributed systems technology X? That may not resonate with management because they don't know or care what distributed systems technology X can do or will do for the organization. Technology X may or may not be the right tool for the job. Instead, you should be talking about what the technology will do for the organization or achieve a particular business result. If you focus on changing a can't to a can, and what the business can do after that, the business will be far more receptive to the new ideas about creating data teams.

How Should the Business Interact with Data Teams?

The business should have consistent and constant interaction with the data teams, just as members of the data teams need high-bandwidth connections among themselves (see the "High-Bandwidth Connections" section in Chapter 7).

It's the data team's job to take the raw data and make data products out of it. It's the business' job to create business value out of the data products—or at least to tell the data team how to make business value out of the data. Sometimes, the data scientist is expected to create this value. Some have a high enough level of domain knowledge to do so, but most don't. You may need a business analyst to augment or fully enable the data scientist's domain knowledge. Different organizations place the dividing line in domain knowledge at different places—mostly in the business analyst or to some degree in the data scientist.

The crucial step of deriving business value from data science is sometimes called BizOps. Some organizations create a BizOps team next to the others described in this book. As I was initially creating the outline for this book, I considered writing an entire chapter to cover BizOps. It's one of those advanced functions that is useful for some companies, but not others. The decision about whether to create a BizOps team comes down to the level of difficulty presented by the domain knowledge of the organization or business unit. The following questions will help you determine the level of complexity in your organization's domain knowledge:

- What is the level of domain knowledge required for a new hire to start being productive?

- Is there a person or persons in the organization whose primary value is knowing how the system works from beginning to end?

- Is there a person or persons in the organization whose primary value is knowing what data is being used or what each field is?

- Is there a complex interaction between third-party data providers and your own system?

- How difficult is it to understand how the data flows through the system from start to finish?

Case Study: Medical Insurance Domain Knowledge

To understand the difficulty and need for domain knowledge, let me describe one of my previous clients. In my experience with clients, the medical insurance industry has some of the most challenging requirements for domain knowledge. Your organization's needs will probably be different, but I want to share the high watermark of what I've dealt with.

Medical insurance companies are hit from several angles with domain knowledge complexity:

Medical

Medical jargon, such as slot, provider, and so on, plus the endless lists of abbreviations.

Insurance

The codings for procedures and the insurance companies' billing processes, which are usually Byzantine.

Billing Systems

The systems where bills are input and then sent to the insurance company.

Data Systems

How the data about the patients, procedures, and bills are stored in the system.

Before I had worked with that medical insurance client, I hadn't fully appreciated the sheer complexity of the domain knowledge needed. I had a conversation with one of their business analysts about where they fit into the data teams and why. At first, I gave the person my standard answer about how the data engineering team should be making data access as simple as possible. The analyst persisted in asking me whether the data engineer or data scientist would even know the right fields to use, which fields to expose, or even what the column names mean.

That was when the light came on for me. Organizations with intricate domain knowledge may need to create a BizOps team that has just as high a bandwidth connection to the other data teams as the three main data teams discussed in this book. The BizOps team needs to be there to provide the domain knowledge and ensure that the data products are usable by the business.

THE RULE, NOT THE EXCEPTION

A message for engineering leaders:

I've learned something over the past four decades or so while working in data and machine learning. Generally speaking when there's an economic crisis, technology applications leap forward—especially in infrastructure. It's almost as if the priorities become suddenly clearer, while the cruft moves out of the way.

Ben Lorica and I spent 2018–2019 surveying about enterprise adoption of "ABC": AI, Big Data, Cloud. We structured our surveys to create a contrast study between firms which already had five or more years recognizing ROI from their investments in this space vs. those which hadn't even getting started. The analysis was compelling, especially when we segmented firms into leaders, followers, and laggards. A small group of "unicorns" of course dominated among machine learning applications, hiring the best talent, using sophisticated technology to leverage their first-mover advantage. Another large segment was following their lead, albeit struggling to staff the appropriate roles, contend with data quality issues, and find enough people who could translate from available technology into business use cases. The remaining organizations—more than half of enterprise—were buried in tech debt, debating with exec staff that did not recognize any need to invest. In other words, that laggard category was years away from having sufficiently effective data infrastructure to be competitive. Moreover, the first movers were aggressively ramping their investments in data, accelerating that gap between "haves" and "have-nots" in enterprise. This analysis correlated with similar studies by MIT Sloan, McKinsey Global Institute, and others.

Looking back at the 2000–2002 economic crisis (Dot Com Bust followed by 9/11) in retrospect the firms that would later take the lead had been busy innovated: Amazon, eBay, Google, and Yahoo had all been working on horizontal scale-out and leveraging machine data by late 1997. In 2001, Leo Breiman published the famous "Two Cultures" article that made the case for industry ML use cases—which before had been rare. In the years that followed, ML built atop big data practices would become the rule, not the exception. None of that would've been thinkable in Fortune 500 companies prior to the Dot Com Bust.

Looking back at the 2008–2009 economic crisis, I'd spent that period building and leading a data team for a thriving second-tier ad network. Just a few years earlier, I'd become a "guinea pig" for a new service called AWS, and convinced our board in August 2006 to let us bet on a 100 percent cloud architecture. A few months into that work, we identified some ML workflow bottlenecks and began using a new open source project called Hadoop to resolve them. Then by mid-2008 we got acqui-hired into the ad network and faced an absurd deadline: five weeks to staff a department of analysts and infrastructure engineers; five weeks to troubleshoot a critical recommender systems based on a failing Netezza instance; and then five weeks to rewrite the firm's main moneymaker as a Hadoop application running on EC2.

Somehow we succeeded. Along the way we hired a contractor named Tom White to fix one of Hadoop's Jira tickets: an I/O problem that prevented efficient jobs on EC2. The result became the largest Hadoop instance running on AWS, and a case study for the Elastic MapReduce service. A year later, Andy Jassy asked me to be a reference customer when he pitched the cloud to SAP, which is one of the more memorable business phone calls of my life. Prior to the 2008 crisis, I could not have imagined making that phone call. On the other side of the crisis, business leaders began referring to this as "data science" teams.

Of course the global crisis in 2020 is substantially more severe than either 2000 or 2008. Nonetheless, those events help us focus on priorities now. It may have been unthinkable in late 2019 that half of the enterprise firms so were buried in tech debt and poor decision-making processes that many would become acquisition targets. Targets by firms that had made investments, cleaned up their tech debt, built effective data infrastructure, staffed data engineering and data science teams appropriately, and embraced practices of leveraging data by teams of people + machines. That's now more likely to become the rule, not the exception. The previously "extreme" strategies now become our ways forward. Given how demands for data analytics have accelerated during this crisis, on the other side of it there will no longer be firms that postpone their decisions to transform and adapt.

So it's a great time to look around. See who's adapting, and how, and with how much success. Also, frankly, see who's not bothering to adapt, and get some distance from them.

The year 2018 and its headlines about GDPR and large-scale security breaches now almost seem like ancient history. Data governance practices have been building, in waves, for the past 30 years or so, but robust DG practices tended to be the exception. Even so, some firms learned from GDPR, although their stories were just barely getting shared before early 2020 hit. I can describe one general set of takeaways, which indicates a general path forward with data engineering become as foundational in business as accounting.

While many firms went begrudgingly into their versions GDPR compliance, one segment realized upside. If you look across Uber, Lyft, Netflix, LinkedIn, Stitch Fix, and other firms roughly that in that level of maturity, they each have an open source project regarding a knowledge graph of metadata about dataset usage—Amundsen, Data Hub, Marquez, and so on.

Collecting that metadata was necessary for them to be prepared to respond to GDRP compliance issues. Organizing that metadata (about how data "links" together) naturally fit into knowledge graphs. Once an organization began to leverage those knowledge graphs, they gained much more than just lineage information. They began to recognize the business process pathways from data collection through data management and into revenue-bearing use cases. They could chart organizational dependencies between data stewardship practices and customer needs. While some identified how their data teams wasted time perpetually "rediscovering" metadata, others recognized possibilities for entirely novel business lines based on their available data.

That was a story just barely getting told in early 2020. It makes the case for having robust data infrastructure as well as robust data governance practices. It converts compliance from a risk to an investment: how does the business overall use its data? Moreover, this approach is purpose-built for leveraging AI to suggest other areas of potential business.

On the other side of this crisis, robust practices for data infrastructure and data governance will be the rule, not the exception.

—Paco Nathan, Author and Evil Mad Scientist, Derwin

Switching from Software as a Product to Data as a Product

Big data and analytics require a switch in mentality that can be difficult for data teams, especially data engineering teams. Software engineering focuses on releasing software. A software engineering team's core value is to release software that fixes a bug or creates a new feature. Data engineering, on the other hand, focuses on creating business value with data. Data engineers are still releasing software. However, they're releasing software that improves the data products.

This shift from software to data as a product can be hard for some organizations and individuals on data teams. Data engineers come from software engineering backgrounds, and your project could be the first time the data engineers have actually created a data product. Management and the business will need to work in concert to make sure the data teams understand that their work focuses on data products and not merely releasing software.

SWITCHING FROM SOFTWARE AS A PRODUCT TO DATA AS A PRODUCT

Deploying software services today involves a fairly mature collection of ideas, like microservice architectures, DevOps practices, use of containers, and so on. Many of these ideas still apply to data-centric products, but there are unique characteristics. Two that are important are (1) the different ways that data scientists vs. data engineers approach their work and (2) the different kinds of application architectures, deployment artifacts, and monitoring requirements for data-centric applications.

Data scientists, like all scientists, use exploration and experimentation to discover the best ways to exploit data, for example, building machine learning models. They are usually not accustomed to the procedural and automation aspects of typical software deployments.

This is really why data engineering exists in the first place, to bridge this gap. Often, data engineers create application infrastructure that provides an easy way for data scientists to add and remove data sources and sinks, and specify new model hyperparameters (model design) and sometimes the model parameters themselves.

However, having data science deliver ad hoc models to production is not recommended. Models are data, so they should be subject to the same governance that other data receives, such as data security, auditing, and traceability. Also, mature production environments automate the creation and management of all artifacts in production, to achieve reproducibility and other goals. Therefore, organizations should actually train models using automated, DevOps-style pipelines, based on the hyperparameters specified by the data scientists.

The production infrastructure will also need to audit and monitor model performance, which records were scored with which model versions, and so on.

A challenge for monitoring model performance is the inherent statistical nature of machine learning models. Engineers are accustomed to expecting exact results (even though distributed systems don't cooperate in this goal). Engineers have to be accustomed to monitoring some metrics that reflect the statistical characteristics of production model serving. Fortunately, data scientists understand probabilities and statistics, so they can help engineers understand how to interpret new kinds of metrics.

—Dean Wampler, Author and Head of Developer Relations, AnyScale

Symptoms of Insufficient or Ineffective Interaction

Often management takes a hands-off approach to their interactions with the data teams. They may work closely only at the start when considering the data strategy and creating the team. But the data strategy is just one of many different phases where interaction is required to achieve value with data.

When the business isn't involved in the entire life cycle of data product creation, the data teams may end up not actually creating business value. The actual data products could be useless or not achieve the business goal. Both the business and the data teams share the blame for these failures. If the data team is using an agile methodology, they should be working directly with the customer—which is often internal business users. Likewise, the business will need to make time or push to have their voices heard during the product life cycle.

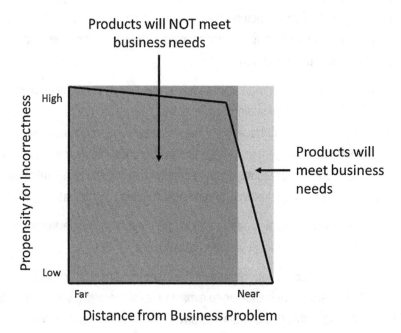

Figure 8-1. *The further the team is from the business problem, the more likely the data products won't match what the customer wanted*

When the data teams are too far away from the business problem, they're likely to create a data project that doesn't meet the business needs. It's only as the data teams get relatively close to the business problem that they're likely to meet the business needs. You'll notice there is a precipitous drop in productive activity as the data teams get really close to the business problem.

I've dealt with teams experiencing the aftermath of a lack of coordination between the data teams and the business. The data teams create what they think solves the problem. From the business point of view, the data product didn't answer enough of the problem to be worthwhile, or it solved the completely wrong problem.

Working with the QA Team

The quality assurance (QA) team is responsible for validating and checking that the software works correctly. For data products, this includes checking performance and the data itself. The QA team confirms that the data products meet the business needs from a quality perspective. This entails checking the software itself, the distributed systems the software runs on, the actual data products being created, and the performance of the software on the system. The team is answering questions such as:

- Does the software work correctly and not throw exceptions?

- Does the model produce the right output given specific input?

- Does the software run correctly on the distributed system?

- Is the data product being created in the correct format?

- Does the software perform at the same level of throughput as the previous version?

Depending on the level of technical abilities of the QA team, these tests can be a mix of automated tests, scripted tests, and manual tests. The more the tests can be automated or scripted, the better off and more consistent the QA testing will be. The QA, data engineering, and data science teams will take care to have sufficient unit test and integration test coverage.

To correctly run performance tests, the QA team will need to have established good baselines against known data input and hardware. Otherwise, any differences in performance could be easily attributed to hardware or data instead of software bugs.

The technical abilities of QA teams vary. Some QA teams do their checks manually, other teams have scripts, and some more advanced QA teams actually write the code to test and automatically validate the code.

QA will need to understand the systems they're validating just as much as the data engineers or data scientists. They may not need to understand it at a depth of the data teams, but it will be essential to know how things should or shouldn't work. The QA team will need to be given similar learning resources as the data teams.

Working with Project Managers

A project manager at an organization is responsible for timelines and calculates the necessary resources to finish a project. A product manager is also responsible for ensuring that technical products can be applied to solve business problems. Often, a product manager is there to conceive of and prioritize new features of a product.

The product managers may work with or come into a data team with small data notions. These small data notions will inform what the product manager thinks is possible or impossible for the technology and data team to accomplish. Product managers will need to be shown what is possible with distributed systems and data teams. This should inform new ideas and methods for improving the product itself.

Project managers also need to understand the increased difficulty and complexity that comes from working with distributed systems. Without an understanding of the significant increase in complexity, the project managers won't be able to create sane and achievable project plans. Unmanageable timelines put the team behind schedule from the very beginning. This, in turn, causes the individual team members to feel as though they aren't doing an excellent job from the very beginning of the project.

Often, project managers will have experience running projects with different technical teams. This can cause project managers to assume that a big data team runs on a similar timeline to a small data team. For example, they may be confused about why a web development team appears to be more productive than the data teams. The management and data teams will want to give the project managers the resources to understand the differences in complexity and timelines when creating data products.

Product managers are marketed to heavily by vendors, who promise the world for what their own products can do. It can be challenging for product managers to sort out what is real from all of the marketing speak. It's also important not to trust all of what vendors say. The data teams—and especially qualified data engineers—need to help product managers sort all of these messages out.

Funding and Resources

Once a business has committed itself to build a data team, management's first task is to give the team the resources it needs, both human and material. Of course, funding will be part of a larger conversation about goals and what's feasible, but we'll start with getting resources in place because nothing can be done without them.

The data teams—or as close to what exists as a data team at that point—will need to create an accurate as possible estimate of the software, hardware, and people costs for creating a data team. This requires buy-in at the highest executive level. As funds for staffing and resources get stuck—and they often do—you need to rely on your executive sponsorship to get things moving again.

Staffing

The growth of data teams requires significant investment in staff, especially at the beginning. Qualified members of data teams—especially data scientist and data engineers—command a premium. The executives and other management should be creating pay grades that fit the market. Otherwise, the executive will have to keep bypassing or doing waivers for the data team member's salaries. Human resources (HR) departments must also be educated about offering reasonable pay grades and pay scales.

HR should get involved in job titles and definitions. Some organizations lack the title data engineer, or their current HR definition of a data engineer is SQL-focused.

The HR team may also have to approve moving managers around to start heading the various teams. They may need to orchestrate the moving of people from another team to the new data team.

Software and Hardware

There is a common misconception among management—especially among C-level executives—that big data is cheap. Many organizations employ free and open source software and derive this low-cost perception from a misunderstanding of what open source means.

Isn't open source free? Yes, it is—with some caveats that you should know about. Let me give you a brief and general overview of open source business models.

Your organization could download the source code from a project's website, compile it, and deploy the resulting binary code. That can be very expensive in staff time and waiting for all the bugs to be ironed out.

A next step up is to take a nicely packaged binary distribution, sometimes already bundled into a virtual machine that was created by a company in the open source space. These binary distributions are often free, too—but they may have some strings attached for how you can use their binary software.

While these binary distributions sound altruistic, they have a hidden agenda. The open source company is trying to make it easy to use their open source distribution so that you can try things out and develop your software against it. Once it's time to deploy things into production, you have to think about service and support for that distribution. Since your code already uses the open source company's distribution, they're the logical choice to call and pay for these services. Although their expertise is valuable, you may find yourself paying more for their support staff than you pay your own staff. Those are backward calculations—you should be investing in your own staff.[1]

Understanding these open source business models, you can see that while open source is free, the associated training, service, and support is not. Some organizations choose to go it alone and try to support everything themselves. This comes with varying degrees of success.

Even so, open source big data alternatives are usually cheaper than closed-source, proprietary options. But the word is *cheaper*, not *free*.

Finally, there are the hardware costs. Whether you're creating a distributed systems cluster on-premises or in the cloud, your hardware costs will go up. This is because many different computers need to be started to run distributed systems and recover from hardware failures. So hardware usage isn't 100 percent. Some distributed filesystems, for example, need 1.5x to 3x as many hard drives to store data redundantly. In order words, to store 10 TB of data, you could need 30 TB of raw storage. This gives some simple examples of how much hardware costs go up with distributed systems and will vary depending on the distributed system itself.

So now we can start talking about getting high-level buy-in. The core reason is cost. It will likely cost more than the executive thinks to create successful data teams.

---

[1]Managers often think they can employ staff with pay below-market rates and make them successful on a big data project by using external service. The results are varied, but, more often than not, these projects fail.

The executive team will need to have bought into the real costs; otherwise, the costs will make any data pipeline project a nonstarter. However, with adequate executive sponsorship, the project and team can have an accurate ballpark estimate for costs.

Cloud

Some management teams and executives think that using the cloud will make big data cheap. Using the cloud can make a move cheaper but not cheap. The data teams can optimize their costs dynamically in a way one can't do with an on-premises cluster.

If the team uses the cloud's managed services, there won't be specific hardware or software costs. There usually aren't any licensing fees or associated costs like an open source vendor. Using the cloud won't make a major difference in team sizes except for the operations team. Using the cloud provider's managed services can make the data engineering and data science teams more productive but usually doesn't dramatically affect team sizes.

The data teams may not be able to make do with only the cloud provider's technologies. They may need to use a vendor's technology. At this point, the team is back to paying for licenses and support with the clusters running on the cloud.

Once again, the cloud provider fees won't be the major cost for the organization. The creation of the data teams and proper staffing will be the major cost. Executive sponsors will need to understand this.

Topics for Interaction

Nearly all organizations should have regular discussions between the data teams and the business about the topics in this section.

Data Strategies

A data strategy consists of looking at the business side to see how data products could augment or automate decision-making. The actual business value can be derived from many places, of which the following are common:

- An internal-facing product to improve marketing or sales through advanced analytics

- An internal-facing product to cut costs

- A customer-facing product

A data product may also serve both internal and external customers.

An entire book could be written on data strategies because they depend on so many different factors: the industry, products, goals, and organization's appetite for risk. I recommend not copying a competitor's data strategy part and parcel because there are profound or subtle nuances that you may not know about.

When creating the data strategy, I recommend that you bring into the room someone knowledgeable about the company's current and potential data sources, as well as someone—the same person, or a different one—who understands the technical possibilities of using the data. This expertise allows the team to toss around ideas without having to continually table discussions for lack of information or ask someone else in the organization the answers to data questions.

Anchoring discussions in technical realities also prevents the team from going too crazy on plans for world domination. Often, teams that are new to distributed systems go too high, too low, or both at the same time concerning what is possible. I've been a part of discussions like these before, and it helps to rein in the silver bullet perception among the nontechnical people that can get out of hand.

A common mistake for management is to think that creating a data strategy is the end of their involvement. Instead, it is just the beginning. Management needs to be there to guide the data teams to a useful business objective. While the data teams will do their best, that may not be enough. The data team needs to have continuous involvement—but not meddling in their technical tasks—from the organization's business-side management.

Risks and Rewards

A crucial part of the data strategy is to evaluate the risk and reward of even doing a project. Creating data teams and data products creates risk. What happens if the project fails? And on the positive side, what happens if the team carries out what they signed up for? What level of value will the team generate for the organization in cost savings or new revenue generation?

When I work with an organization on their data strategy, I look for a value generation of at least 5x to 10x over the costs of the project. This level of value generation may not be possible initially in the "crawl" phase but should be achieved later in the "walk" and "run" phases. Remember that value generation for data team projects may not be seen in increased revenue but in cost savings.

If a project fails to create a considerable amount of value, the knee-jerk reaction is to blame the technology. Instead, the management team should really double-check their motivations for data projects. It could be that the management team didn't exhaustively find all value that could be generated through data, or it could mean that the organization isn't thinking about their data as an asset that could be leveraged. The issue could stem from an improperly resourced data team.

In this case, "do it right" means possessing a data team with the right skills and enough resources to accomplish the task you have assigned. Putting your team in an untenable position is unfair and creates morale problems. These morale problems eventually lead to people leaving the team or organization.

Projects that fail for these reasons start out with uninformed mandates from upper management. Someone at the top found out about distributed systems and wants them at their company, but didn't consider the requirements described in this book.

A project with these starting points represents the highest risk of failures, with the lowest value generation. Although a book can't offer specific advice to avoid such an outcome, I highly recommend reevaluating the risk/reward of these projects.

Managing and Creating Realistic Goals

Perhaps the most challenging upper management issue is creating realistic expectations and goals for the data teams. Executives are often targeted aggressively by marketing hype around the possibilities of data products and the value they create. This marketing, in turn, can be communicated to the rest of the organization by the other executives who have second-hand or third-hand knowledge of what's possible. These activities can lead to an entire department or organization with unrealistic goals and ideas of what is possible.

A data team that faces unrealistic goals from the start has a crippling problem. They will fall behind or end up unable to deliver. I highly encourage managing these expectations as early in the process as possible. Remember that you're not just fighting what's being said in your organization—you're fighting what the various vendors' salespeople and marketing machines are telling them what's possible. All of these people leave out key details and caveats that need to be managed.

Use Cases and Technology Choices

Use cases, which I've mentioned from time to time in this book, are crucial to your success with distributed systems.

With small data and nondistributed systems, 99.9 percent of projects could use the same technology stack. For me personally, that meant that I did 99.9 percent of my work with Java, a relational database like MySQL, and a web application technology like Apache Tomcat. When I went into a meeting to discuss a new feature, there was rarely a discussion around which technologies we might need or could improve the project. It was always around how to use our existing technologies differently or improve our usage.

With distributed systems, you have to deeply understand the use cases before even talking about which technologies to use.[2] This is because distributed systems have trade-offs that aren't present with small data. The ways in which these systems have to scale and the different parts of the system that risk becoming bottlenecks determine how the particular distributed system cheats. I'm not talking about cheating in a negative sense. These cheats are necessary to achieve the scales that big data provides. Without them, you just can't scale. See "The Role of Architects" section in Chapter 4 and my blog posting.[3]

Making technology choices before fully understanding the use case leaves you with the probability that the technologies may not be able to handle the use case. You may choose the wrong tool for the job, and that could mean that it takes an hour to do something that the use case expects to take a second. It could mean that the technology can do only batch processing, or that the technology takes at least 1000 ms to process input, whereas the use case requires processing in real time at less than 200 ms.

Understanding the use cases before choosing technologies is one of the first things I try to teach new data engineering teams and new data engineers. When I do an architecture review, we spend hours on the use case. We go through every nook and cranny of what they want to do and the business reasons for doing it. Often, the data engineers are getting anxious to start whiteboarding and diagramming out the architecture. I press them to get deeper into the use case and the whys behind it. Only then do we start talking about technology. This way, we really know what we need to before we even start thinking about technical architecture or implementation.

---

[2]Read more at my blog posting This is Useless (Without Use Cases) (`https://www.jesse-anderson.com/2017/07/this-is-useless-without-use-cases/`).

[3]Read more at my blog posting On Cheating with Big Data (`www.jesse-anderson.com/2017/10/on-cheating-with-big-data/`).

Speaking Truth to Power

A mid-stage to late-stage issue for upper management comes from the data itself. Sometimes your organization's data or analytics will fly in the face of an executive's decision or strategy. The data team needs to be able to speak truth to power without the fear of being fired. The fear of losing projects or being fired leads the data teams to withhold the full story from executives. The worst outcome is that the teams will start changing their findings to agree with the executive's preconceived notions. Being a data-driven company isn't about politics; it's about improving the business and decisions.

Executives may have skewed views based on previous experience that wasn't data-augmented. A model's predictions or analysis may go against their previous experience. The data scientists and executives may have to work together to dispel these notions and work through trusting the results. Trust in results can be built up by showing the hypothesis that were either proven or disproven and the work that went into creating these results.

This tolerance for bad news grows with the organization's maturity. A data-augmented organization will learn or know how to deal with and use data to its maximum value. Organizations that are just learning how to use data when making a decision will need a feeling-out period.

Speaking truth to power doesn't mean the data team should be belligerent. That does happen sometimes and should be noticed and discouraged.

Executive-Level Attention

The majority of the day-to-day work is done by the organization's middle management, but executives are an essential contributor to the success of data teams. Earlier in the "Staffing" section of this chapter, we saw that they need to speak up early on to fund the organization's data team. These decisions should be based on the data strategy and on continually refined estimates of the value the organization's data can create.

Some changes required for data teams have to happen at the top. This can happen because there is a need to coordinate and work with other parts of the organization. I've seen middle management fight to push changes upward, a Herculean task. These changes are much better and more efficient when done at the top.

A later decision revolves around the project's success. If progress stalls or a project fails, the executive-level management will decide whether to keep it going.

Dealing with Uncertainty

Data science is mostly a research endeavor. The actual outcome of a data science project is uncertain. Will it even work? Will they also find a problem that is solvable with an algorithm?

This stands in contrast to most engineering tasks. Given some engineering tasks, there is a 99.9 percent chance of the engineers finding something to improve or do with the data. With data science, there isn't a 99.9 percent chance of a positive outcome. A perfectly viable result is that nothing comes out of the time the data scientist spends on the task. The possible uncertain results are something that management often doesn't know or understand.

This uncertainty isn't just for the outcome; it extends to the amount of time needed to vet the viability. This is what makes data science so difficult to run because you can't run it like an engineering team. You have to run it more like a research and development team.

A significant exception to this uncertainty is when the use case or data is clearly established to have applications in data science. Some of these clear-cut examples are fraud detection in financial use cases or recommendation engines in various industries. We know it's possible to use the data for these use cases. The big question is how good the results are and the impacts on the business.

Data Sources Are Changing

Added to this difficulty is the changing nature of the data. This is even worse with high periodicity to data or when there is little historical data. This means that the data for models changes often, but you don't have enough historical data to see the periodic trends.

Even in well-established companies and industries, the data will change. I've seen this happen for different reasons, but I'll share an easy-to-understand example. In finance, it is prevalent to use machine learning in fraud detection. On the other side of these models are the people trying to perpetuate the fraud. These fraudsters don't just wait around to see if they get lucky with fraud. They're continually seeking new and more advanced ways to defraud others. The data science team will need to keep abreast of the data source changes—be they on purpose or by chance.

Other times we need to be able to reproduce a model's results. This can be due to a regulatory environment or an actual law. You may need to prove or show how a model scored a piece of data. You'll need various piece data that can change over time. You'd need to know the exact code used to do the training and scoring and the correct data that was used to train the model. Some organizations think this can be done with timestamps and backtracking, but that isn't always possible, especially with the data sources changing. Some organizations embed this metadata in every message to understand what is happening in the system during the scoring.

Output of Models Is a Level of Certainty

The output of a model for scoring or inference isn't a true/false field. It's actually a certainty or confidence interval. For people who aren't used to data science, this comes as a real shock—especially for engineers. The model gives you a positive or negative certainty (usually a floating-point number) of what it thinks is happening. The final consumers of this data product will have to decide the levels or values they're expecting to be normal or within their tolerance.

Don't Fear the Reaper's Data Products

The data scientist's analysis and results—in an ironic twist—are often unused or ignored. This can be because the organization doesn't believe the results, fears them, or doesn't understand the math behind them. Finally, of course, they can't act because the results go against the highest paid opinion.[4] It's sad when an organization goes to the effort and expense of creating a data science organization and proceeds to ignore its output.

This is a juncture where the root of the problem isn't technology or data science. The issue is purely a human one—with the lack of trust being the core of the problem. It's a lack of trust in the person or team who created the algorithm. The solution is to start getting your organization comfortable with trusting an algorithm and the team that created it.

This problem is especially prevalent in old-school organizations. These are organizations where an executive's intuition was the driving force behind decisions. For these sorts of organizations, I recommend positioning the model or output as augmenting the decision or decision-maker rather than replacing them altogether.

---

[4]This is a common colloquialism in business. Sometimes, decisions aren't based on the data or results. They're based on the opinion of whatever person in the room makes the most money—and therefore must be recognized as the smartest—instead of what the person who is most expert in the subject says.

Although this isn't the optimal scenario for the data science team, it gets their data product into usage and gives humans the time and space to catch up.

Don't Forget the People Side

Business changes—particularly those created by data teams—could affect people throughout the organization. These people are equally critical to the success or failure of projects because their enthusiasm or resistance can affect the data team positively or negatively.

When I work with an organization, I don't just ask questions about the business value or technical issues—I ask about the politics and buy-in within the organization. This must be considered throughout the company in order to prevent all of the data team's hard work from going nowhere.

For instance, does the sales team know what impact you want the data to have on their work, and have you consulted with them on the proposed optimizations to sales leads? Do the customer service representatives know how you're going to optimize call routing based on their specializations and training? Does the field sales team understand the changes and benefits you hope to bring to lead routing?

If you don't take the human side into account, you can have a mutiny on your hands. The revolt is entirely preventable and comes down to communication. Communicate what is happening and why. Win people over to your side, or they may sabotage the data team. There are many ways to do this: failing to provide input, ignoring recommendations, or even actively working to make the project fail.

Data Warehousing/DBA Staff

One of the teams facing major changes with the advent of modern data teams are the data warehousing teams and DBAs. The new data teams are taking on and replacing the use cases currently being handled by these teams.

Frankly, most organizations have been abusing databases for something they weren't good at. For example, large-scale analytical queries that have to scan the entire dataset aren't good uses for relational databases. Once you remove that abuse and use other technologies, you are better off. Better off means that the tinkering and constant problems that kept data warehouse teams busy to go away. And that, in turn, means your DBA team doesn't need to be as large.

The management team will need to take a look at the direct effects on the nondata teams and make a 6-month to 12-month assessment of them. They'll be faced with some difficult questions about workforce planning. How many people will need to be on the other teams once the data teams are handling their use cases? How many can upgrade to data engineers and be retrained? How many can find places elsewhere in the company? Do they need to find positions outside the company?

Don't get me wrong—organizations with DBAs and data warehouse teams aren't going away entirely. However, their number of people will gradually go down and won't be backfilled. That said, organizations that never had data warehouse teams and that are creating new big data teams don't go back and start data warehouse teams.

SQL Developers/ETL Developers

SQL and ETL developers are a tricky group to handle too. More often, the use cases are too complicated for their technical abilities. They have a challenging time becoming data engineers. The majority of the time, their programming skills aren't ready and will take a significant amount of time to get prepared. From the organization's point of view, you may always need some SQL queries written, but your level of demand may not remain high enough to keep the ETL or SQL development teams at their current headcounts.

Some people debate whether SQL is a programming language. Immaterial of that, the move from SQL to Java isn't a lateral movement like C# to Java. Between SQL and Java, there is a big jump in syntactic complexity and object-oriented concepts. This translates into learning from scratch, more than an application of previous knowledge. In contrast, a move from C# to Java is more a translation of "I already know how to do this in C#, and I just need to figure out how to do it in Java."

SQL-focused staff may be able to write a query to power a report. Just know that they may depend on others for advanced programming help and can create these data products only from a data product produced by the data engineering team. For some organizations, there is enough need for reports to keep SQL and ETL staff busy. For other organizations, the user—often a business user—is expected to write their own SQL and, therefore, to self-serve. It's at this point that you need even fewer SQL-focused people.

SQL is great for certain things, but when someone is limited to only SQL, it becomes abused. I've seen massive and inefficient queries written in SQL that abuse the language. The query worked adequately but was incredibly brittle. Only the person who wrote the query originally could understand it, and it would have been better as a combination of SQL with code in some traditional programming language such as C.

Operations

Changes to headcounts for operations will depend on whether the cluster will be on-premises or in the cloud. It will also depend on whether the data engineering team will be practicing DevOps.

Cloud services, especially managed services, need fewer personnel devoted to operations. The day-to-day hardware issues will mostly be handled by the cloud provider. With managed services, the framework operations will also primarily be handled by the cloud provider.

Not every operations person can learn distributed systems. The operations team will need to become skilled at troubleshooting and operating each distributed system. This push to specialize further into distributed systems may create the need for a separate operations team that deals only with distributed systems.

If the organization chooses to practice DevOps, the operations team may need to integrate within the data engineering team or figure out how to triage issues as they arise.

Business Intelligence and Data Analyst Teams

In general, the headcount for business intelligence or data analyst teams will change very little, if at all, during the move to big data. With data teams, the business intelligence and data analyst teams are getting new data products for analysis, not being taken over.

Some organizations promote or move some of their more technically proficient data analysts to the title of data scientists or to be part of the data science team. If these people aren't 100 percent at the required level of technical or mathematical level, the organization should be sure to provide the necessary learning resources.

Establishing Key Performance Indicators (KPIs)

Creating KPIs for data teams gives them a specific direction and goals to accomplish. Other than company-wide KPIs or targets, I believe all data team KPIs should revolve around the question, "Are you creating a data product that creates business value?" If a data team isn't creating a usable data product, there should be a KPI that management can point to that they aren't fulfilling.

Here are a few suggestions of KPIs for each team.

Data Science Team

- Increase in automation of a manual task

- Decreased time, cost, or fraud

- Increase in sales, conversions, or widgets

- Decreased turnaround time for an analytic or machine learning algorithm

Data Engineering Team

- Improvement in data quality

- Socialization of data internally and/or externally

- Increased self-service of data for internal consumption

- (For new data engineering teams) Enabling the data science team to do previously impossible tasks

- Improvements to data scientist quality of work life

- Increased automation of machine learning, code deployment, or code creation

Operations

- Decreased downtime

- Decreased meantime to restoration of service

- Increased security of data and data access

- Increased automation of deployments

- Increased captured and monitoring of metrics

CHAPTER 9

Managing Big Data Projects

Loving you whether, whether
Times are good or bad, happy or sad

—"Let's Stay Together" by Al Green

In earlier chapters, we've staffed your data teams and found support for them within the larger organization. This chapter covers a number of day-to-day and long-term issues that managers have to face as the teams progress:

- Planning the creation and use of data products

- Assigning tasks to the right team

- The special needs of data scientists

- Long-term project management

- Technology choices

- When it all goes wrong

- Unexpected consequences

Planning the Creation and Use of Data Products

The value delivered by data teams comes in the form of its data products, which vary across a range of lifetimes. Some may be single, one-off reports, whereas others are exposed through convenient channels where business users call on the data products over and over during decision-making.

© Jesse Anderson 2020
J. Anderson, *Data Teams*, https://doi.org/10.1007/978-1-4842-6228-3_9

Sometimes, the data engineering team gets things started by creating new or improving existing data products. The data science team consumes those data products. As part of that consumption, the data science team may create derivative data products. Finally, the operations teams keep everything running correctly and optimally so that the data products continue to be generated automatically.

One-Off and Ad Hoc Insights

Data engineering usually focuses on creating data pipelines that are continually in production and with a long-term view. This is how most data engineering is done, but not every data product needs this level of rigor.

One-off or ad hoc insight analysis doesn't need to have the same level of engineering because it doesn't need a long-term view. Instead, it would be slowed down by too much engineering effort. Whenever possible, the ad hoc insights should use existing data products and infrastructure that is already in place. If ad hoc insights are becoming more common and needing a faster turnaround time, the team should look at creating a DataOps team. This could make the business happier with faster turnaround times.

In these scenarios, the team's success or KPIs should be measured in the team's ability to execute quickly and their ability to use existing resources as much as possible. As you find a resource that is consistently being used but isn't exposed as a usable data product, the data engineering team may need to put the engineering effort into exposing the data pipeline continuously and automatically. From there, the ad hoc efforts can leverage the new data pipeline more effectively.

At another, higher level of usage, when you notice a number of similar ad hoc usages, it's time for the data engineering team to put into production a consistent analytic pipeline or report.

Finally, management should be aware of what fast and ad hoc means. It means that the team doesn't put the usual engineering rigor into an insight. Instead, the team focuses on speed over engineering. This difference in speed can cause other managers to wonder why everything isn't this fast. They may also misunderstand when some insight is production-worthy or is just an ad hoc insight. They may think an insight is ready for long-term usage when it isn't. The data team should take care to help the organization's management to know when a data product is or isn't production-worthy.

Exposing Data Products

Letting your business know how to use the data team's data products is key to adoption and usage. The actual data products should be exposed with the right technologies. Notice that "technologies" here is plural. This is because distributed systems will cheat in various ways (see "The Role of Architects" section in Chapter 4).

The actual technologies used to expose data products depend entirely on the use case. For example, queries that run quickly and return a small amount of data could be exposed with a RESTful call.[1] Meanwhile, queries that operate on and return large amounts of data need to have the data exposed directly through the distributed system. Qualified data engineers and architects will make sure that the data products are exposed the right way.

Often, management will want data products to be exposed with a single technology. This isn't always possible due to a use case's specific read and write patterns.

A particular distributed system's cheat will make it an excellent choice for one technology and a terrible choice for another. This forces the data engineering team—in concert with the business side—to stand up a new distributed system to handle use cases that need access through a different technology.

Assigning Tasks to Teams

Although Part 2 tried to draw clean distinctions between data science, data engineering, operations, and other related teams, real life is messier. Sometimes when a new project starts, someone must determine which tasks go with which team.

Management will be the first level of triage when looking at a task, project, or use case. The management team will have to decide which team or teams are needed to work on the project. Getting this right is essential so that the right teams are engaged from the outset of the project instead of finding out too late that they need help from another team. Data teams need to validate they are the right team for the job and can fulfill the requests. When the wrong choice is made up-front, one team could have to play catch-up or redo the work from another team completely.

---

[1]A RESTful call is a web-based API that is exposed with HTTP verbs to interact with data. For example, a team could expose a REST call with a GET verb to retrieve data about a customer.

Data Engineering, Data Science, or Operations?

When choosing teams to route a task to, think about the issue that is at the root use case. Of course, decision-makers must understand what each team really does and what their core competency is.

The data engineering team is right for engineering tasks. Examples of suitable tasks for data engineering include one or more of the following traits:

- Heavy on code development

- Optimizing an algorithm, instead of creating one from scratch

- Making a report or analytic go faster by changing the underlying technology or stack

The data science team is right for advanced analytics and machine learning tasks, characterized by one or more of the following:

- Math-focused or statistics-focused

- Optimizing a business approach or outcome

- Improving the quality or statistics behind an analytic or report

The operations team is good for things that are already in production or are ready to be put in production, with one or more of the following traits:

- A simple task that could be scripted

- Automating an existing pipeline or code sample

- A data product already in production that is misbehaving

Collaboration

Some tasks may need all three teams at once. This is when things get really complicated because you have to break up what is ostensibly an easy task into multiple and independent tasks. From there, you have to coordinate different parts of the organization in separate teams. Finally, the independent tasks come back together into an interdependent set of tasks that all work together.

As you receive a task, make sure to include an estimate of the time spent on sheer overhead to coordinate a task that is spread across all three teams. This overhead includes the meetings that have to be held, which in turn include preparation for the meetings, the actual time spent in the meetings, and the increased nodes of communication that have to be created. The members of the data teams need to all be in sync for the task to be completed correctly.

Rectifying Problems with Existing Teams

Sometimes your data teams aren't effective. They aren't working well together, maintaining high-bandwidth connections, and creating value with data. When a problem happens, there's lots of finger-pointing and no real resolution. The business isn't able to use the data products they're created. In such cases, the primary task of the organization is to fix the data teams.

When I work with organizations, I spend just as much time talking to the business side as I do talking to the technical side. By talking to the business side, I speak to the people who are the actual end users of the data and thus are constantly interacting with the data.

Talking to the consumers of data reveals the problems on the data teams. If the data science team doesn't work with the data engineering team, you will hear it in the low value of insights created. If the data engineering team isn't producing good data products, you will see that the business doesn't trust the results. If the operations team isn't keeping things running in production, you will hear about the constant production outages that prevent the business from really trusting that the data project will be working when they need it.

Once you've seen the reflected problems, you can confirm their existence with the team. And after confirming the problems, you can start to get to the root of them and fix them.

The source of problems with existing teams can be many and varied. Teams made up of great individuals, both management and individual contributors, can still miscommunicate and create poor results. On the other hand, teams made up of not-so-great individuals may not be pulling their weight. It takes some frank looks at the team to figure out the real root cause and not just the superficial issue.

Effects of Using the Wrong Team for a Task

The implications of choosing the wrong team for a task are significantly increased by the complexity and coordination required for distributed systems.

A poor choice that is common, particularly among new managers, is to assign the data science team to tasks that should be done by data engineers. This choice might be made because there's no data engineering team, or because it is too new within the organization. The person routing the tasks might just fail to understand the difference between these teams.

To give you an idea of the implications of using data science for what would be a data engineering task, let me share a scenario I've seen in many different organizations. The symptom of the problem is data scientists that are either stuck or making glacial progress on a task. The manager will ask me to talk to the data scientist about what is happening. Usually, the data scientist has been at the task for a month or more and has made virtually no headway.

What we find is that the data scientist was tackling a task that's 99 percent data engineering. By data engineering standards, the task would be relatively easy and doable within a few days. But the data scientists lack the technical ability or knowledge to choose the right tool, approach, or algorithm for the job. Instead, the data scientist is trying to use what they've always been using to solve the problem. When a project is blocked, it rarely works to double down on the previous approach; the team should look for another method instead.

There are counterexamples where data science tasks were assigned to a data engineer. It just doesn't turn out well.

Long-Term Project Management

When data teams have long-term projects similar to software engineering, organizations need to put a general road map in place. This is because data teams need to coordinate with other teams, departments, and business units. The road map could include the operationalization of new technologies, the creation of new data products, or improvements to existing data products. Without a good plan, the data teams will be purely reactive in creating and implementing data products.

Some teams try to plan too far out into the future. Near-term plans—about one year out—should be relatively concrete, calling out specifics, whereas longer-term plans, more than a year out, should be much looser and be more conceptual.

This is because your data team should be changing in a year. These changes could be both positive and negative. An example of a positive change could have the data teams improve their velocity (see the "Data Project Velocity" section in Chapter 7), so that a project that was planned for six months now takes only three. An example of a negative change could be the loss of a project veteran or another key member of the team. Now a project that was going to be assigned to the project veteran and take three months would be assigned to several junior members of the team and take nine months.

Really long-term plans risk ignoring the speed at which distributed systems are evolving. Planning three years out could have drastic technical implications for a data team. A newer technology, or a newer version of the technology the organization is using, may make you rethink how you implement a feature. And this change, in turn, could affect how long the team needs to implement the feature. This makes the team have to come up with a new plan or amount of time to implement the feature.

Technology Choices

Managers are not expected to know the technologies used by data teams, which are not only complicated but in constant flux. This section is directed to managers to help them vet technologies and direct their data teams toward choices that help the entire organization. In some sections, I talk about particular technologies (notably programming languages), but I explain the key concepts managers need.

A Mental Framework for Adding or Choosing Technologies

Management can help data teams choose the right technologies by using the mental framework I lay out in this section. Technologies will come and go, but this framework will be the same.

When an organization is deciding to add new technology, I suggest doing a risk/reward calculation. This includes establishing the clear business value of adding new technology. The risk/reward calculation will help to answer any pushback from management about the cost of adding the new technology. By setting the business value, the data teams should be able to succinctly answer what they can enable within the company by adding a technology.

Keep in mind that not all technologies directly add business value; some are more foundational. These foundational technologies will enable future use cases that are anticipated or will allow faster or easier implementation of new features.

Management should understand the long-term effects of adding new technology. Is this technology or implementation a hideous workaround that we'll live to regret?

This may take some really pointed and challenging talks with the architect or engineer proposing the technology. Remember that there are often many different ways of solving a project. Management will want to know what other possibilities were explored and the reasons for choosing the proposed route. While management may not be technically trained enough to understand or honestly critique the choices, they can try to understand the motivation for the decision.

Beware of the promise that new technology will make everything easier. This promise could come from the team itself or the vendor that is pushing it. Adding new technology could make certain parts easier while making other elements much more difficult. For other technologies, you may not see the benefit for months or years after implementation. In my opinion, distributed systems can't be made easy. Some parts can be made easier, but the improvement is limited. When I start to hear a vendor tell me that their distributed system is easy, my ears perk up and suggest checking what else they're lying about or over-promising.

Thus, managers can sometimes head off wrong choices. When the organization is hiring an architect or lead engineer, the potential hire should be asked about their propensity to choose new technologies. Ask them what happened when they added a new technology or design. Did it really do what they thought it would do? If not, how did they react? Did they seek to cover over the problem and shift blame, or had they tried to mitigate their own and the organization's own risks? An organization should try to avoid hiring people that strive to add complexity to a system—either knowingly or unknowingly—because they may be adding complexity for all the wrong reasons.

Reining in Technology Choices

Managers often push back when the purchase of technologies is requested from them. They ask whether data teams really need 10 to 30 different technologies. Surely, they think the data engineering team is over-engineering a solution or padding their resumes. They consider the cost of operationalizing each one of these technologies and decide they aren't all really needed for a data product.

A second typical pushback I get is whether the distributed systems technologies are really as complicated as I make them out to be. This feeling can spring from comparing the big data team to other software engineering or data warehouse teams within the organization, which superficially appear to be more productive while doing the same tasks. This perception wraps back to the previous management misconception. The data teams are interacting with—and actually do need—some 10–30 different technologies to be successful, whereas the software engineering and data warehouse teams are interacting with one to three technologies. Management often asks the follow-up question of why can't data teams just use one to three technologies or merge their requested 10–30 different technologies into a few.

The reply stems from the observations I made in the "Use Cases and Technology Choices" section in Chapter 8. When data teams expose data products, they're optimizing for a particular use case. A new technology that isn't already by part of your ecosystem might be a much better choice for the use case. By focusing on enabling the use case, the data team will have to choose the right tool for the job, and that may mean operationalizing a new technology.

As management, it is permissible to push back on operationalizing a new technology. This is because data teams may not always have the best or most altruistic motives for choosing a new technology. These self-serving technology choices could revolve around being bored with the technologies the organization is currently using. It could be that a data engineer feels like their resume is becoming stagnant and isn't filled with the latest and greatest technologies. They're getting ready to leave the organization and need to make sure their resume shows that they're doing new and exciting things.

Most of the time, however, their reason is to use the right tool for the job.

In any case, there is an inherent cost to operationalizing a new technology. When adding new technology, there should be a clear and beneficial business case that the new technology allows or accelerates.

LOVING TO HATE TECHNOLOGY

Many people have told me of their stress in selecting a technology for their future projects. I think the main problem with this task is that it includes so many unknowns. There are the unknowns of the requirements and scale of the future project. Then there are the unknowns about newer technologies that you may not have firsthand experience in. On top of that, we are bombarded by advertisements for the latest and coolest technology, and in fact, we are rewarded financially when we select technology that become newly marketable. So what to do and where to begin?

The following advice is for one that wants to be in successful projects while also positioning themselves for the future. Learn to only select technology that you hate and look to hate more technology every day. That may first come off as odd advice but listen for a minute.

When you learn more about a tech, you learn about what it can do and what it can't. I find that the best technology selection is when you know fully well what is bad about a technology but still it is the best technology for the task. If you follow this paradigm, you will never fall into the following pitfalls:

- The over promised tech

- The tech that requires super high-tech talent to manage

- The tech that solves for problems you don't have

—Ted Malaska, Author, Director of Enterprise Architecture, Capital One

Programming Language Support

Enterprises and even small organizations may use several different programming languages. While some organizations try to have a single language, this isn't always possible because of the mix of data science teams and data engineering teams. Other organizations relinquish all control over the choice of programming languages to the data teams and hope they choose the right tool for the job. This can lead to many, many different languages being used in production systems. Because data is being exposed, you will need to deal with exposing data products into each one of these languages.

Exposing data products—as opposed to an API endpoint—is one of the key differentiators between data products and straight software products. Management and technical decision-makers will need to factor in how to expose data products to each language as they choose whether to support that language.

Support for a distributed system varies significantly among different programming languages. Some languages have great first-class support, whereas others have quasi-first-class support. In the second case, the language is only partially up to date but lags behind the latest features.

Still other languages offer support as a separate open source project. For example, the distributed system could exist as an Apache Foundation project and support a single language. Someone outside of the project will see the need or have the desire to support their language of choice. They will create their own implementation to support that distributed system.

Often, these other language support implementations will be hosted on GitHub. Not all of these implementations are created equal. Some are started out of a single person's passion, whereas others are created by a company to support their own needs. I generally recommend that organizations go with the implementations that are in production at a company and are generating value for established companies. These projects are more likely to get the continued care and attention they need.

Languages Used by Data Scientists

Data scientists primarily use Python, but some use Scala, which is compatible with Java libraries. The vast majority of products aimed at data scientists will expose a Python API, perhaps along with other languages.

THE VALUE OF THE JVM

A bit of technical background is useful here.

Java is converted (compiled) to a fast-running format that runs inside a framework called the Java virtual machine (JVM). Many other languages were designed to be compiled to the same JVM, so they can work easily with Java and take full advantage of the many powerful libraries created in Java. Besides Java, Scala is the most popular language based on the JVM. Scala has some more modern programming language features than Java and takes up less space, so it's usually faster to code.

The data science team is a consumer of data products created by the data engineering team. The data engineering has gone through the effort of exposing data projects using the right technologies. This helps to mitigate the limitations of non-Java language support. The newer or full features for the language's support may not be needed because the data engineers have exposed it properly.

Languages Used by Data Engineers

Data engineers primarily use Java for their work. Some teams will use Scala, usually along with Java. And some use Python.

As the creator of the data products, the data engineers bear the technical brunt. They face the most technically challenging part of data teams' responsibility to create and expose the data products. This technical challenge requires that data engineers have as many tools at their disposal as possible to correctly expose data products. The choice of language directly affects the ability even to choose a technology because that technology

145

may not have an API supported in the team's language of choice. Missing support for your language of choice could mean that a use case isn't possible or that the team will have to wait until the distributed system starts supporting the language. That could take months or years for both closed source and open source technologies.

Most distributed systems—notably big data technologies—target and support Java first. This is because the technology or distributed system itself is programmed to run on a JVM (see the sidebar "The Value of the JVM" earlier in this chapter).

Exposing Data Across Languages

There are several issues with exposing data products across languages. Which format should you use? How will you account for the differences in typing systems between the languages? Is the type a floating point or an integer? What is the size of the integer: 32-bit or 64-bit? Is the integer signed or unsigned?

Each one of these questions might sound like computer science topics that don't really make a difference, but they do. An incorrect type could mean the difference between a correct count and a wrong count for a report. Figuring out the source of an inaccurate report could take days and require some deep dives into the guts of how a programming language works. This is where an ounce of prevention will be worth many pounds of cure.

The architects and data engineers will take care to have well-thought-out answers to these questions before they come up. Binary formats like Google's Protobuf and Apache Avro are often used to solve these problems. Please take this section as a strong recommendation to get this crucial part of exposing data products right.

Project Management

For project management, data teams usually use either Kanban or some version of Scrum. Kanban and Scrum can also be used together.

The choice of project management framework depends partly on the estimated length of time a project needs. The needs of data engineering teams can be similar to or slightly different from those of a traditional software engineering team. For more conventional software engineering tasks, a Scrum-style framework can work well. A more conventional task is a long-term project similar to traditional software projects.

For other tasks and projects that have to be completed more quickly, a Scrum-style framework can be too much overhead. This will decrease the responsiveness and increase time spent on reporting on short-lived tasks. For this type of operation, Kanban may be the right project management framework.

Data science teams have specific and unique needs for project management frameworks. For a data science team, reading an academic paper in preparation for a project doesn't fit well into a Scrum. The team's focus on experimentation, with the possibility or potential for disproving a hypothesis, doesn't lend itself well to Scrum styles. This leaves a data science team with the choice of Kanban or another low-to-no overhead-style project management framework. Some data science teams create their own project management framework that fits their working style.

The operations team may favor a Kanban project management framework. This allows the operations team to keep track of and prioritize any noncritical work.

I have found some teams still using waterfall or a similar derivative, and I suggest moving to an Agile framework. This is because data engineering teams—especially new data engineering teams—need to start showing value on a shorter-term basis. Operating on waterfall timelines prevents the team from demonstrating value or creating data products until years pass.

If a data science team doesn't use Scrum, how should they integrate back into or work alongside a data engineering team that is using Scrum? For example, a data engineering team could be working on 2-week sprints, while the data science team is working on open-ended or time-boxed projects.

In such a situation, some organizations will wait until the data scientists are done with something and then schedule it into their sprint. Others will keep the data engineering team at a lower capacity to have time to interact with the data science team. For example, if a data engineering team would generally be 90 percent loaded, the manager would lower that to 70 percent or some other appropriate number to allow for extra time to work with the data scientists.

When It All Goes Wrong

Sometimes projects will go off the rails, despite your best efforts as a manager. It takes extra initiative to figure out and fix the source—or usually several sources—of the problem. Just the task of figuring out the issue can be incredibly difficult. In my experience, issues in data teams don't rectify themselves without concerted effort.

You may have seen error bars on charts before. These show the margin of error that was measured when the measurements or data was taken.

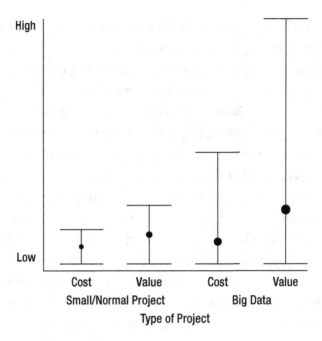

Figure 9-1. *Big data projects have a much higher risk/reward than small data*

With small data, a project's error bars are relatively small. Even so, when a small data project goes off the rails, it can take weeks, months, or a year to fix. With distributed systems, a project's error bars are much more significant. That means that the project could fail more miserably than you've ever experienced. (On the positive side, it could succeed and generate far more value than you could imagine.) A big data project going off the rails will take many months or years to fix.

It's only with proper planning, the right team, and the right resources that a team has a fighting chance at succeeding. Management needs to keep a consistent watch for the signs that a project is going off track. In my experience, it's never too late to save a project. The older a project is, the more expensive it will be to save it, both in time and money. That said, if a company is looking at shutting down their big data projects, it's probably because so much time has gone by, and so little value has been generated. I suggest using this book as a guide to figure out what needs to be done to get the projects making progress again. This will give management the specific steps and changes that can be shared with those management team members that are in favor of shutting down the project.

Look Out for Nth Order Consequences

Usually, the true cause of a problem is hidden, and what you see is a third-, fourth-, or fifth-order consequence. I can't go through every single manifestation of issues due to time and space constraints. Just know that there are further problems down the line as the team progresses and increases their velocity. In this book, I'm focusing on the first- and second-order consequences and the most common long-term consequences. Some advanced or older teams will face additional issues that I don't address in this book.

Just to give you a taste, let me share a few examples:

- How do you handle the departure of your team's veteran? They probably possess much of the folklore of how things work. How do you handle being left with mostly—if not exclusively—beginners who were really relying on the veteran to figure things out? Worse yet, what do you do if the project veteran was really the only one who was productive?

- What should you do if your team is so successful that other organizations start to poach your staff? How do you keep you people while other organizations are offering them a hefty pay raise that your organization won't provide?

These are just a few examples of the third-, fourth-, and fifth-order consequences of success, or sometimes a mix of success and failure. Some issues can be industry-specific, a historical reason within the company, or a purely political issue. The more specific the problem, the less apparent it could be to inside observers, and the more difficult it will be to fix.

CHAPTER 10

Starting a Team

We'll find a place where there's room to grow
And yes, we've just begun

—"We've Only Just Begun" by Carpenters

In earlier chapters, we've talked about the teams and how to manage them. This chapter covers the steps and considerations for starting and hiring the first members of the team. For organizations that have already started hiring, this will be a reminder of what should have been done or confirmation that the right decisions were made.

- Making sure the team starts out properly

- Deciding where to locate the teams in the organization

- The considerations of what to do with the previous teams

- Getting the team right depending on the size of the organization

- What should be the reporting structure of the data teams

- Working with human resources on titles and pay

Establishing New Teams

Some organizations will need to establish a data team from scratch. This includes hiring or choosing a manager to run the team. Then you have to hire the individual contributors on the teams. This is an excellent opportunity for the organization to get their teams and management right from the beginning.

As you plan a timeline for hirings, make sure that you keep in mind the lead times for the entire process. You may want to ask the HR department how long it takes from the time the headcount is approved all the way until someone is sitting in a seat. The HR department may not have any data on hiring a data team member, but you can use

151

© Jesse Anderson 2020
J. Anderson, *Data Teams*, https://doi.org/10.1007/978-1-4842-6228-3_10

similar titles as a touchpoint for how long it takes. For example, if the average hiring time is three months for a software engineer, you can assume it will take at least three months to hire a data engineer—if not more.

Failing to take into account the lead times for hiring can put the team behind in a project plan from the very beginning. Also, waiting until the team is actually ready or needing the new team member puts you behind. Ideally, the team guesses how soon a new team member is needed, and the hiring process starts before then, leaving the average number of months it takes to make a hire.

The First Data Engineer

Because data engineers are the creators of data products, I recommend hiring data engineers first. They need to start getting everything ready for the subsequent data team members. As you will see in Chapter 11, you need a qualified data engineer in the presteps. This is because only a qualified data engineer can make some of the critical early calls on what data teams should and shouldn't do.

The first data engineer has to start getting the infrastructure and data products in place. Up to this point, other data team members would be mostly idle, waiting for data products. The lone or small group of data engineers may need several iterations before the next hire should be made.

DON'T HIRE DATA SCIENTIST FIRST

A common mistake when creating a data team is to hire the data scientists first. This comes from a management and HR misunderstanding of the differences between a data scientist and data engineer.

When hiring from scratch, there is a chicken-and-egg problem. As you've read, the data teams are very tightly integrated, and you need all three teams. But when you're starting to hire, you have to start with one person. Some organizations hire the data scientist, thinking that the best value is to start getting some analytics. The issue with this thinking is that data scientists often lack the data engineering skills to even get enough data together to begin analyzing it. As I've mentioned, data scientists are the consumer of the data products that data engineers create.

In addition, many data scientists really hate doing data engineering work. There is really a different mindset between the two groups. Some data scientists will join a team expecting that the organization will hire a data engineer shortly after. If an organization doesn't hire a data engineer, the data scientist may quit after six months or so.

The First Data Scientist

A data scientist should be hired once the organization has their data pipelines and data infrastructure in place. Until this time, the data scientist time is mostly idle while waiting for data products and the infrastructure with which to test hypotheses. This is because the data scientists are the consumers of data products, not the creators of data products.

Operations

Before going into production with a data pipeline, the organization will need to hire operations engineers. For some organizations and use cases, an operations person may need to be hired sooner. This can be the case when the system being created is quite complicated, and an operations engineer will require more time to get familiar with the data pipeline.

Some organizations practice DevOps and have their data engineering team handle the operations of the distributed systems. The management team will need to make sure that the data engineering team is given the resources to learn how to operate the distributed systems they are expected to maintain.

If an organization doesn't practice DevOps and still expects its data engineering team to handle all support, several problems could spring up. Some data engineers really don't like doing operational tasks and will quit. Other times, the data engineers aren't the best operational support because there is a different mindset for supporting systems that some data engineers lack. A data engineer who isn't careful in supporting a production system can cause data loss. Also, the recurring pressures of handling operations take the engineers away from their primary job of creating data pipelines. If this happens often enough, the data engineers will be unable to make progress.

In any scenario, even with cloud usage, people will need to support and operate the cluster. For some organizations, this can involve a 24/7 support service-level agreement. Professional operations staff are the only ones who can take on such responsibility and should be expected to do so.

As operations team members are hired, management should expect the operations engineers to reduce the operational and support load on the data engineers. For a brand-new data pipeline, the initial support from the operations may be relegated to more general operations. As time goes by, the operations team should be gaining more experience and taking on a more significant support load for the other data teams.

The Critical First Hire

The first person hired for the overall data team and the first within a team is a critical hire. At this point, there isn't anyone to give pushback on a bad idea, and this first hire makes all the decisions. This one-person show plays an essential role in the success or failure of the data products. Your role as a manager is to find the right person for the job.

The first hire sets the stage for everyone that comes after and lays the early foundation for the data products. Their decisions—whether good, bad, or a mix of both—will affect everyone who comes after. Poor choices can have exceptionally long-lasting effects.

Bad decisions, architecture problems, and poor technology choices will box you into a corner. The use of distributed systems magnifies problems. That means that you can't just take a week to rewrite some code or fix the architecture. Often, rewrites of code and architectural changes can take months or an entire year.

The first person also sets the technical culture of the team, repelling or attracting future candidates and employees. The first hire determines core issues such as the following:

- Is it a team that fears critiques and feedback?

- Is it a team that focuses on best practices and iteration?

- Is their design so terrible that new job candidates don't want to deal with it?

- Is their design so good that the job candidate realizes they can learn from the team?

As you can see, a good data engineering culture will bring in and attract the right people. From the outside, the candidates will perceive the best practices being used and want to be part of the team. They'll see that they can learn and improve with the team. Additionally, a good data engineering culture will screen out those who don't want to adhere to best practices.

A bad or poor data engineering culture may turn away good candidates. From the outside, they can see turmoil and problems created by poor technology choices and problematic architectures.

Location and Status of the Data Team

The day-to-day and organizational structure either contributes to or reduces the effectiveness of the data teams.

Embedding the Data Team in Business Unit

A common decision is whether to embed the members of a data team within the business unit. Executives sometimes do this when a business unit feels that their use case or requirements aren't getting enough attention. They'll think that the only way to get someone's attention is by having their own dedicated person. This complaint could come from a place of real need. It could be the result of a data team that isn't prioritizing correctly or fairly. In any case, the decision requires some serious considerations due to the ramifications to the data teams.

One consideration is that moving data engineers and data scientists outside of the core team could make them bored or unhappy. The business unit's use case could have been boring in the first place, and that might be why no one wanted to work on it. Boring projects lead to unhappy people, and unhappy people tend to look for employment outside or the team or organization. The unit where you embed people could turn into a place where people go to quit the organization.

To prevent boredom and unhappiness, the organization may need to take steps to head off the issues. This could include rotating team members in and out of a business unit. Rotating will at least give the team members a deadline, after which they'll switch back into the central team, and maybe something more interesting.

The critical variable determining whether rotation will work is the level of domain knowledge that a business unit needs. If the business unit has relatively low domain knowledge requirements, the people can rotate easier and more quickly. If the business unit has high domain knowledge requirements, there will be more difficulty in rotating people in and out. This is a place where management really needs to find a good trade-off between ramping up on domain knowledge and keeping people from quitting.

Another method is for business unit staff to have a dotted-line relationship with a member or members of the data teams. This dotted line means that the person remains a member of the data team but recognizes that their rewards are based on meeting the needs of the business unit. This helps the data team members deal with the loneliness that can start when someone is doing data team problems but aren't around other data team members to keep their sanity.

A big issue that plagues business units that hire their own people is that the hiring managers know little to nothing about hiring data scientists or data engineers. This lack of knowledge usually condemns the business unit to employing the wrong person or an unqualified person. In turn, this puts further strain on the data teams to support the business unit and the person who hired to do the job in a problematic and strained attempt to get things done. Further on in time, the decentralized data scientists or data engineers aren't exposed to the organization's best practices or tooling and will suffer from inconsistent training and career development.

Hub and Spoke

Another variation of working inside an organization is the hub and spoke model. This model features a centralized hub team. As the team gets big enough, several members of the team can move off and start a spoke that is directly in the business unit.

The benefit of this model is that the splintered team members will still have close ties to the hub team and can continue to enforce the use of best practices.

Center of Excellence

Some organizations choose to create a center of excellence. Depending on the size of the organization, this center of excellence will be part of the enterprise-wide team, or the center of excellence could be located in a business unit. This center of excellence team is tasked with creating and modeling the best practices around the organization.

For the center of excellence to work, managers need to make sure that the team is genuinely modeling and using best practices. Otherwise, the center of excellence will spread the issues and dysfunction that cause teams within the organization to be underperforming or failing.

Data Democratization

Still, other organizations go a route of just making the data available to the business unit. This approach is often called *data democratization* or *self-service.* The organization creates data and the infrastructure on which to run analytics, but sometimes no one has a specific mandate to do the analytics. Some organizations do a mix, where some teams have a mandate for analysis and other teams can do analysis as they have the need.

When making data available, remember that when everyone is responsible for a task, no one does the task. Every team has enough to do already. Adding one more optional task will go nowhere unless the teams really have a dire need for analysis.

This approach also tends to underestimate the complexity of doing analysis. Even when the data engineering team does a great job exposing the data products, the complexity of performing the analysis can still creep in. This can mean there are many different people all trying to do analysis, and they're unable to do it.

Sometimes, the real crux for data democratization is the organization's lack of resources—particularly in education and learning—that enable business units to self-serve or not. Without these resources, the rest of the organization could write poorly performing code or not be able to perform the analysis at all.

Data democratization also raises security concerns when the organization deals with sensitive data. Care should be taken to verify and validate the effectiveness of the organization's security policies to keep sensitive data safe.

SUCCESSFUL DATA DEMOCRATIZATION

Organizations that adopted Hadoop early tended to adopt data democratization, almost by accident, due to the limited set of features in early Hadoop versions. The cost of operating a Hadoop cluster gave incentive toward having a single large cluster with all data, and weak support for access control resulted in data being widely accessible. Moreover, the lack of support for transactions and mutable data storage forced adopters into workflows based on immutable datasets and data pipelines. Immutable data can be shared with little operational risk, so friction for data reuse is minimal in a processing environment based on pipelines. The early Hadoop adopters were fast-growing and frequently changing organizations, which were comfortable adopting new workflows. As a result, they effectively broke down data silos, and data from one team could be used for innovation throughout organizations. Enterprise Hadoop adoption has not had the same democratic effect; the enterprises are less amenable to change workflows, and Hadoop vendors have added better support for fine-grained access control and data mutability in response, ironically diluting the disruptive effect of Hadoop.

In an organization where data is not democratized, getting data access can be challenging. A feature team with an innovative idea that requires data from systems owned by other teams needs attention from those teams. They need support to extract data, documentation to understand it, assistance to transform it from the source system's internal structure, and operational collaboration in order to avoid disrupting production source systems. For all of these steps, the source system teams need to allocate time in the backlog. This synchronization of backlogs and priorities adds friction to data-driven innovation. A feature team operating in an organization where data, including metadata and documentation, is democratized can put data-driven features in production faster.

The data platform team is key to successful data democratization. Their incentives and definition of success should be aligned with analytics and feature teams' definitions of success while managing global risk. The data platform team's product manager has the delicate task of balancing the desires of feature teams to use data without friction while maintaining necessary governance and security practices. There is a risk of strain between platform and feature teams, in particular if they are geographically separated. Moving individuals between teams has proven to be effective for overcoming such strains. Embedding platform team members with their internal customers provides efficient feedback to platform development. Likewise, letting feature team members have a temporary assignment building platform features disseminates understanding of platform constraints.

—Lars Albertsson, Founder, Mapflat

Getting the Team Right from the Start

Earlier chapters of this book described the three main teams every big data project should have, and the basic mix of skills needed. Beginning a team presents special challenges that will be described for different types of organizations.

Data Science Startups

When your business is dedicated to data science, either as a consultancy or in service of a single project such as drug development, you'll feel pressure to hit the ground running and do some data science right out of the gate.

Unless you're fortunate, you will still need all three data teams described in Part 2. A team consisting only of data scientists will get themselves into trouble on the data engineering parts and create technical debt that needs to be paid down.

Companies Lacking a Data Focus

Some organizations are incredibly good at their domain—manufacturing, retail, health—but not as good on the IT side. Even working with small data, they're barely getting by or are scarcely competent. Organizations like these may not be able to succeed at marshaling and deriving insights from data because it isn't their forte.

Consider whether your company has the management skills for data and IT and whether you are ready for the major changes to become a technology-focused company. You should definitely try to use the benefits of analytics, but perhaps by contracting out to a contractor whose expertise is there.

And be warned: the really hard thing about benefitting from big data isn't collecting and deriving analytics. The hard thing is figuring out how analytics can promote meaningful behavior change in your organization, and having the fortitude to follow through and make the changes that your analysts recommend.

Some organizations are putting faith in a false positive that they're ready to start their own data teams. This false positive could be that the organization is using an off-the-shelf program to do their analytics, or they are leveraging a consultancy. An organization could be a very advanced user of an off-the-shelf program but could be nowhere near the necessary technical levels to create their own data products. When an organization uses a consultancy, the consultants could be doing the really hard parts of the implementation, and the organization's own people are really coasting behind their hard work. Organizations that don't take an honest look at their core competencies could find out the hard way the complexities that were being done for them.

Small Organizations and Startups

In small organizations, each hire is a major decision because they have so few opportunities to add staff. The first hire in a data team reflects where the organization thinks the most significant or best investment lies.

Small organizations often lack the resources to hire all of the needed members of a data team. When one person has to handle the tasks of all three data teams, the hiring managers have to settle for someone who has uneven skills in all those areas and must decide what strengths are most important.

I recommend that people in this situation find a data engineer with some knowledge of analytics. If there is a business intelligence team, the data engineer could start working with the business intelligence team to start creating analytics with the data products the data engineer has created. During this time, the data engineer should be set up for the next stage when you hire data scientists and other staff.

You may also find that the analytics you need aren't as complicated as you initially thought and therefore can be generated by the data engineer. It's also useful to keep the data engineer busy managing your data if you have to delay or find another way to start using that nifty machine learning model that you promised your investors.

If you hire a data scientist first, just remember that there will be a period where even the more technical data scientists will be creating technical debt (see the "Technical Debt in Data Science Teams" section in Chapter 3), which you'll have to pay down later. Expect this technical debt to mount between the time you hire your data scientist and the time you hire your first data engineer.

Also, you probably haven't hired an operations person yet. If you're lucky, your data engineer will be somewhat versed in DevOps. This will force you to reduce your operational load as much as possible, probably by making extensive use of managed services on the cloud. As I've mentioned in the "Cloud vs. On-Premises Systems" section in Chapter 5, these services don't completely negate the need for operations, but they do reduce the operational overhead significantly. Yes, managed services are more expensive than a do-it-yourself route, but the cost advantage of doing it in-house will quickly be swallowed up by the first significant and time-consuming outage.

Also remember that the more time your person is spending on operations, the less code they're writing.

Consequences of a Bad Hire

Small teams find themselves in a precarious position because they have only enough budget for one person and that one person is responsible for hiring subsequent team members.

If you hire someone who doesn't understand the requirements of data science or the needs of your business, the subsequent hires will replicate that wrong decision several times over. The first person you hire may actually reject excellent candidates who might show them up or perform much better. Their architectural and design decisions might repel the good hires because no one wants to spend their days cleaning up someone's dumpster fire or find themselves fighting their own manager from the start.

Thus, for a beginning team, this first person is a crucial hire. Fixing the aftermath of this decision could take months or years.

This brings up the question of whether hiring a reasonable-looking candidate, even if you have doubts about their skills, is better than delaying the start of a data team. In my experience, it pays to wait and find the right person.

Consequences of Hiring Only Junior People

Some organizations try to work around the issues of hiring correctly by hiring several junior people instead of a top-notch, experienced person. The organization's logic is that the skills of the team will average out and that one member can plug any holes left by another member. In any case, there will be plenty of workers to write any code. If there's a problem, there will be a large group of people to work on it.

The reality is that a group of junior people doesn't possess complementary skills. They will all have the same beginner understandings of what to do. Managers imagine a pie chart with ten empty spaces showing the necessary skills for a data team (Figure 10-1). In their mind, the junior people will scatter about and fit nicely into all ten empty slots, with just bits of space in between where the person doesn't have 100 percent of that skill. The imaginary pie chart is 90–95 percent filled. But that's not the reality. Actually, the pie chart is only 10 percent filled, because the junior people's skills and abilities are all stacked, one on top of another.

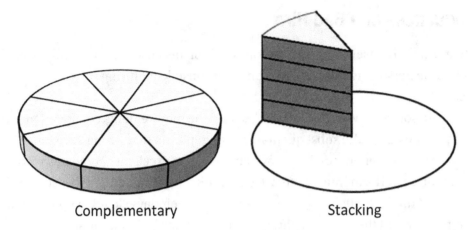

Complementary   Stacking

Figure 10-1. *Beginners stack on top of each other, and correct teams complement each other*

Usually, success on these teams is the result of sheer luck, like looking at a stopped clock that's right twice a day. You may end up finding a junior person with tremendous potential who pulls above their weight. But that's unlikely—better to spring the cash for an experienced person and fill the organizational chart in later with junior people that the experienced one can train.

Medium-Sized Organizations

This is an organization that already has an IT or software engineering function. Some of these software engineers in the IT department could have the makings to become data engineers. Some software engineers may even have done some data engineering projects. On the data science side, there should be an analytics or business intelligence team already at the company.

It's natural to look at the staff you trust, and whose skills you know, to stock your data team. But beware of stirring up disrupting existing corporate functions and internal politics. Your software engineering or business intelligence team will be losing a member. Often, these staff are senior members of the team, which can represent a sizable loss to the team. The people who are moving also have existing projects and responsibilities. Make sure the person isn't torn between their new data engineering team and their previous software engineering team. The person won't be able to make progress if they're constantly interrupted with prior responsibilities.

Sometimes management at a medium-sized organization can have sizable ambitions that far outstrip their budget and team's ability to produce. These organizations tend to get bogged down on some copying some cutting-edge company's architecture or idea heard at a conference talk.[1] These projects and teams tend to go nowhere and make no progress on things that matter. It may be possible for an organization to replicate the cutting-edge company's abilities, but this could take a long time to accomplish. These long time frames aren't exciting or create enough value, and the projects get canceled.

Instead of replicating something that seems to have worked for companies, look for projects that create unique value from data for your organization. How can your organization and customers benefit from data? Appropriate projects are exciting to executives and get increased funding. In contrast, projects that get stuck or produce less value than the cost of the team tend to cut or canceled.

In general, medium-sized organizations need to be informed about where to spend their resources. They're just big enough to be too ambitious. They're also small enough to really feel the pain of a data project going nowhere. Data projects done the right way can really be a catalyst to impressive growth in a medium-sized organization.

Large Organizations

Large organizations usually have extra income to invest in data teams and attract good candidates, so the main problems there tend to be political. These politics can arise from all types of antecedents like merging companies, different chains of command or departments, and national or cultural differences.

Whatever the root cause for politics, the manifestation is that data gets siloed into many different pieces. You may hear your data analysts or data scientists complain that they can't get all of the data they need from a team or that it takes too long to get all of the data together into a single place. As you approach these projects, make sure you understand that there is both a technical problem (assembling and harmonizing many

---

[1]Sometimes medium-sized organizations will send their top people to a conference in the hopes of finding cheap advice and companies' architecture to copy. They'll seek out a company doing cutting-edge things with data and seek to copy them. They won't know just how difficult and costly it is to do cutting-edge things. Other talks will cover a new technical idea or concept. These talks often come from or are promoted by a vendor whose technology facilitates that idea. The idea could be rooted in something proven in the real world, or it could be the company trying to do a technical land grab of something that isn't proven or, worst case, a really bad idea.

different datasets) and a management problem (dealing with different organizational cultures and competition). If you leave it up to your individual contributors to fight these political battles day after day, people are likely to leave.

Likewise, a large organization's slow crawl to fix identified problems or make any other change can be tedious to data teams. This lack of movement can kill any progress and give the team a feeling of being stuck despite all of their efforts. At these large organizations, you may be forced to work with departments that have never worked with another one or are dreaded for their less than stellar performance. And yet, you will have to figure out a way to work with them.

Another problem can be caused if an organization goes from no data team to 20–30 people very quickly. This can lead to a deadlock, where all 30 people are idling and waiting for data or for a person to give them something. In particular, data scientists tend to sit idle while waiting for data and infrastructure to work with. This puts the data engineering team under higher pressure to create what they need—often putting in more hours than average—while the data scientists complain of nothing to do. The lesson: don't make the team big until you've achieved your fundamentals, which consists of the technology and datasets the data scientists can use.

At large organizations, sometimes an individual contributor's greatest asset is their knowledge of the organization's quirks and history. I've heard these people referred to as data archaeologists. You may need to have these people on the team to unearth or relate what happened. These people are a valued part of a data team's domain knowledge.

What Should the Reporting Structure Look Like?

The reporting structure of data teams is essential to how the business interacts with the data teams. The correct reporting structure can lead to organizational alignment and common goals. The wrong reporting structure can lead to friction and nonoverlapping or conflicting goals.

Under the right leader, data teams can become more productive. It can be challenging to find the right leadership for the data team because not every department lends itself well to the activities and roles of data teams.

Let's consider some possible leaders to lead all of the data teams. Depending on the size of the organization, the data teams may report to a person under that executive's department.

VP of Engineering

A VP of Engineering with a competent mathematical background can be a good choice. Their experience with math may let them be satisfied with the uncertainty inherent to data science. The VP of Engineering may have had enough exposure to statistics to understand some of the math and concepts used by data scientists at a high level.

Familiarity with statistics is useful not just so they understand the statistical models (at that high level, such understanding may not be necessary), but to hold the right expectations for data science projects in general. The goal of a project is often to disprove a hypothesis. The team may report, after spending months looking at the problem, that a project or idea isn't possible. This situation doesn't exist in software engineering and could be a rude awakening for leadership.

A VP of Engineering who lacks statistical knowledge may be the right manager for data engineering and operations, but not for data science.

Some people argue that the data engineering team absolutely has to be part of the larger engineering organization. Otherwise, data engineering will be dismissed as not being real engineering. And in turn, the rest of the organization might not treat the data engineering team as having the same software engineering pedigree as the other software engineers. This can lead to infighting and a dearth of the cooperation that the data engineers need with the software engineering teams.

CTO

The CTO may not have a background very similar to the background described earlier for a VP of Engineering. The CTO could have come up through the ranks in the same way and may have even held the title of VP of Engineering. Just like the VP of Engineering, the decision comes down to how comfortable they are with math. If they're not comfortable with the statistics, this may be the right place for data engineering and operations, but not for data science.

CEO

The CEO has lots of ideas, which are a good thing if they're executed correctly. But lots of ideas can mean the CEO is driven by the wind. This means that data teams under the CEO could be given a brand-new priority every day. This clearly isn't a good position to put the team into.

If a team is given a new priority that takes weeks or months to accomplish, the team will have to take a least a day to plan it out. If the CEO proceeds to give another task that takes weeks or months to accomplish, the team will spend the next day just planning again. After enough times of getting brand-new tasks every day, the team will spend all of their time planning and not executing.

This absurd sequence eventually gets to the point where the team will just stop planning and just sit around waiting—knowing that any time spent designing or implementing a project will just be wasted. At this point, the team will just start quitting.

CIO

The title and duties of the CIO can be highly varied. Choosing the CIO as the leader for data teams really depends on the organization's definition of CIO. Some CIOs may have software engineering teams or analytics teams working under them. These CIOs could be a great choice, given that they're already familiar with both the software engineering and analytics requirements. Other CIOs have a different definition that is more focused on IT spending and operations. This may not be the right place for data teams.

VP of Marketing

The VP of Marketing may not be the best manager for data teams. They can be too focused on key performance indicators (KPIs) for data science to have the room for proper experimentation, hypothesizing, and failure. Each organization's marketing department can be different, but marketing and data engineering may not mix well either.

VP of Product

The VP of Product may not be the best place for data teams. Similar to the VP of Marketing, they can be too focused on KPIs. They also can be too focused on creating a product instead of the data that can improve other parts of the business. This can lead to contention and fighting over the resources that will cut costs or otherwise help a business be more efficient.

The VP of Product is another job title embodying a lot of variety. In some organizations, the software engineering and operations teams report to them. This could be the right place for the data engineering and operations teams, but care should be taken.

VP of Finance

Some people highly recommend the VP of Finance as the place for data science. This is because the VP of Finance will have a good background on the math side. They may even be comfortable with the uncertainty of statistical models if they've studied that branch of math.

In my experience, the VP of Finance will push back too much on costs. While they may recognize the value and inherent cost of hiring a data science team, they may not understand the need for all three teams to be successful. This difficulty can be due to their not realizing the complexity inherent to data engineering and operations.

Chief Data Officer (CDO)

The CDO is an emerging role that some organizations choose as the home for their data teams. Like any emerging title or position, the CDO definition can vary significantly in organizations. At some organizations, it can be a technical person who came up through technical management, or it can be a person with little to no experience with data. Some CDOs will have little to no experience managing a data team. Others can come from highly regulated industries where data strategies revolve around defensive strategies and mindsets that could inhibit creating value from data.

VP of Data

The VP of Data is another emerging role. This can be the right place for the data science team.

Depending on the VP's background, they could have extensive experience in software or data engineering. But often, the person came up through the data scientist ranks and lacks this background in software or data engineering. This could make the person unfamiliar with, or fail to enforce, the engineering rigors necessary to be successful.

Working with Human Resources Departments

Some organizational changes need to happen on organization charts and on paper rather than with code and technology. This means involving an organization's human resources (HR) department. Your work with HR may range from a quick chat to establish job descriptions to a journal through a labyrinthine maze of the enterprise HR. In any case, the HR department needs to be part of the conversation around the changes required for the data teams.

The issues with HR are part of the reason management needs C-level buy-in. You may get pushback from HR about the whys, whats, and hows of creating the data team. The recruiting part of HR could be stymied by the differences in recruiting and hiring data teams when their usual tactics for hiring don't work. You may need C-level pressure to push through an HR holdup.

This Isn't Just a Focus on Titles

One typical pushback is around title changes. The HR department may question the need for new titles. For example, they may ask why a data engineer is any different from a software engineer. They may ask why a data scientist is any different from a data analyst or business intelligence analyst.

A part of HR's purview is to define what a title does and what an employee gets paid. Thus, your discussion is about the job description and pay, more than just titles.

Specializations such as data engineers or data scientists expect higher compensation than a person without a specialty. In fact, I would argue that data engineers and data scientists are sub-subspecializations. For example, a data engineer is someone who has specialized in distributed systems, software engineering, and technology. It's because of these specializations that a data engineer likely will not accept the pay scale of a software engineer.

Some organizations will try to offer the high end of a pay scale for a software engineer, but that rarely works because even the high end isn't high enough. Other organizations will try to one-off their pay offers for data team members and handle each salary negotiation separately. This is slow and occupies more time from executives than it should. As a direct result of too stringent guidelines on pay, I've had my clients miss out on outstanding candidates.

This focus on pay doesn't mean these data team members are money hungry. The individuals have chosen to further specialize in order to have a higher demand for their skills and, therefore, a higher pay scale. For data scientists, they're facing the prospect of paying back the student loans for their Ph.D. From a purely monetary perspective, the nonspecialized salaries are often 20–40 percent or more under what the market rate is compared to the real specialty salaries.

Adding New Titles

At most organizations, it's the HR department's job to bless and add new job titles to the recognized list of job classifications. Convincing HR that a new specialization is actually required may take time. I would put this book forward as a means of convincing HR and showing the differences between the specialized job descriptions and the less skilled job descriptions.

Some organizations may already have the titles for the data team members. After really understanding what each title really does, you may need to double-check that the HR job description really matches the reality. If it doesn't, you will need to work with the HR department to revise the job description to a more correct definition. Having an incorrect job description for the title will make the wrong people apply for open job positions. Some organizations have further HR paperwork and descriptions that you will need to seek out and possibly correct.

Title of Data Scientist

Adding the title of a data scientist to a blessed list of job descriptions is usually one of the easier tasks, due to the buzz around being a data scientist. Many HR people have read about it. The HR people may already be seeking to add the title due to the external and internal pressure to start doing data science.

If the title exists already, you will need to ensure that the title is defined correctly. This will include day-to-day tasks. The job description should also specify the real technical expectations for the data scientists and the languages the team is using.

Review the requirements for a data scientist. If a data scientist job description requires a Ph.D., does this really reflect what the job requires or the education level of the rest of the team? Some universities do have data science programs, but they aren't prevalent. There are boot camps and other non-Ph.D. programs that can produce competent data scientists. The job description will need to be clear whether these boot camp graduates, Ph.D.-level graduates without a specialization in data science, or people without a Ph.D. will be accepted.

At some organizations, there is a push to further break down the title of a data scientist into more granular parts. Some data scientists may focus more on the research side or into the application of data science to business problems. Others may focus on a specific technical domain such as natural language processing (NLP) or computer vision. Some organizations list such specialties with different job titles from that of a data scientist.

Title of Data Engineer

Adding the title of a data engineer can be more difficult. From an HR perspective, the difference between a software engineer and a data engineer seems like splitting hairs. It may be challenging to explain the increase in complexity between the distributed systems a data engineer deals with and the small data systems a software engineer works with. Care should be taken to explain the reasons for needing the title of a data engineer.

If the title exists already, make sure that the job description is defined correctly. Some definitions of data engineer focus on small data and SQL-based systems instead of the distributed systems and strong programming background needed by a data engineer, as defined in this book.

Few universities offer data engineering as a specialty. Some data engineers come from boot camps, while still others are self-taught on distributed systems and data engineering. Make sure that the job requirements state the level of experience the organization is willing to accept.

Title of Operations Engineer

Adding the title—or more correctly, the specialization within operations—of operations engineer may be the most difficult to convince HR. The HR department may find it really difficult to understand how difficult it is to troubleshoot and maintain running distributed systems. HR may try to say that any operations engineer can handle the operations of a cluster.

If the title exists, make sure the job description adequately calls out the distributed systems technologies that the operations engineers will be expected to support. The major of operations engineers in data teams will have come from an operations background and just started to specialize in supporting distributed systems.

Focus on Technology Skills

I find that HR job descriptions and screenings focus too much on technology buzzwords. Does the person have X distributed systems technology on their resume? HR assumes that if the technology isn't listed, the candidate must be unable to perform the job. The reality is that the answer is more nuanced.

HR should know that moving from small data technologies to distributed systems is a big move. Distributed systems are far more complicated. The person has to deal with data in a way that small data systems don't. It isn't an impossible move by any means, and the HR team should take that into account.

The move from one distributed system to another distributed system is much less complicated. For example, if the candidate knows a distributed system that is similar to the required distributed system, chances are the person will have a much easier time learning the new distributed system. The learning will more entail understanding the differences between the two distributed systems and learning the new API. In this sense, moving from one distributed system to another is more of a lateral move.

Many job candidates have no professional experience with distributed systems. This is going to be the norm for years to come because there are few universities offering courses on distributed systems. New hires will have the difficulty of wanting to get into distributed systems with no experience.

To start off with, professional experience in distributed systems doesn't directly correlate with actual skill. Some resumes look fantastic, and yet the person isn't remotely qualified. It is possible to learn distributed systems technologies without having real-world experience. It's still up to the candidate to prove that they can create systems with distributed systems, and I recommend these candidates do a personal project to show their skills.[2]

In evaluating people without distributed systems experience, you may want to ask some follow-up questions. Has the candidate done anything similar to distributed systems before? Are they bored with small data technologies, and are these skills too easy for them? These are some consistent markers I've seen for people who are ready to make the switch from small data to distributed systems.

---

[2]See Chapter 7 "How Do You Get a Job?" of my *Ultimate Guide to Switch Careers to Big Data* book to see more discussion about personal projects and their usage in showing competency and mastery.

CHAPTER 11

The Steps for Successful Big Data Projects

South Central does it like nobody does
This is how we do it

> —"This Is How We Do It" by Montell Jordan

Running big data projects is different from running small data programming or analytics projects. Beyond obvious tasks such as choosing datasets and technologies, managers and data teams need to put many subtle building blocks in place. Some of this work may be done before the data team is even hired. The steps for each data project are laid out in this chapter, roughly in the order in which they should occur.

The steps in this chapter are the most universal, applicable to virtually every project and organization. Other steps may be specific to certain industries, to organizations of certain sizes, or to particular organizations.

Presteps

These steps should be taken before the first line of code is written, and even before making any technology choices. Successful projects are successful well before the first line of code is written. They start with proper planning and focus.

The management team is the primary team accomplishing these presteps. However, they will need the help of a qualified data engineer or data architect. Getting the input of technical people this early could pose problems because these steps could come so early in the process that no one on the data team may be hired yet. It could be that you have only a junior data engineer to consult when a qualified data engineer is required for these steps. Depending on the size of the organization, there may be another team with a data architect or qualified data engineer available. Some organizations look outside for help. Getting accurate and competent help at this stage will really pay huge dividends down the line.

Some data teams are tempted to skip the prework and go straight into starting the project by choosing technologies and starting to code. If a data team skips these preparatory steps, the team risks going idle and making no progress while someone hurriedly tries to finish the prework. These are expensive mistakes.

Look at Data and Data Problems

Because the results of analytics depend first and foremost on the organization's data, management will need to go through the organization to find where there are problems with data. This will involve talking to the business and other nontechnical users or potential users of data. The best sources of information are the more technical users. These conversations will reveal the available data, what shape it is in, and who owns it. It will also provide an idea of how siloed the data is and how political the fight will be to get ahold of the data.

During this stage, ask users what they can't do that they'd like to do. These "can'ts" should be technical; otherwise, the data teams may not be able to solve them. These can'ts should be things like

- It takes too long for the report/analytic/etc. to run.

- Whenever I ask for a report to be run, I'm told that it will bring down the production database.

- The team tells me my use case isn't possible, but I have the sneaking suspicion that is possible—just not for them.

Some questions the management team may want to ask are

- What can't you do now?

- If you could do that, what value would you add to the organization?

- If you were to put a monetary value on the can'ts, how much do you think they would be worth to fix?

- Why do you think it can't be done?

Once you have a good idea of your data and the politics around it, you can start to form a data strategy. This is where the input of expert technical staff becomes crucial. Without it, management will create a plan that's unfeasible or too complicated. I find that upper management tends to create a strategy that's too ethereal or buzzwordy to carry out. To prevent that problem, I suggest creating short-term or medium-term data strategies with focused and attainable goals. Management should really be more focused on an iterative plan that is updated regularly as the team and data products move along.

Identify Business Needs

All data projects should have a clear business need. Some data projects are foundational, started in the hope of enabling future improvements in the business. But the next step after a foundational data project should have a clear business need. Just spinning up clusters of distributed systems is of no value until there is a clear vision of how they're going to be used and benefit the business.

Establishing business needs or value can be difficult for managers who are new to running data teams. Management may not be able to ask directly what their business need would be and implement it. It could be that the business side lacks the technical knowledge to answer that question. They may not know the value of data or what data could do for their business yet. Or, the business ask could be so technology specific and prescriptive that the solution only solves a specific tactical issue instead of a broader set of needs.

When neither side knows or can articulate what needs to be done, the data team enters a deadlock. This sometimes happens when the business expects the data teams to be the drivers of all things data and technical. When this happens, I suggest having a workshop with both the business side and the technical side to talk about what could be done. There needs to be a qualified data engineer at this meeting to make sure what is discussed is feasible.

The workshop won't be able to answer the direct question of what data products need to be created, so you'll have to focus on the secondary effects of data value. How much money could the organization save if you were to exploit data better? How much money could the organization make, or how much could it optimize the sales process? Could the business increase customer goodwill by improving customer engagement via the website, email, or in-person interactions?

By asking secondary effect questions and having a good dialog about what to do, the organization can start to articulate a clear business need. Data teams without a precise business need have a challenging time creating value from data.

Is the Problem a Big Data Problem?

Another vital question to answer is whether the data or use cases involve big data. The usual definition of "big data" doesn't specify a particular size, but defines it as data too big to process using current systems. This is really a two-step question. Do you have big data now? Will it be big data in the future?

If you don't have a big data problem, maybe a different team is better suited to handle the job, such as a data warehouse team or a software engineering team. If it is a big data problem, you can start building a case for data teams.

As you decide whether a use case is big data, make sure you have access to someone technically competent who validates the decision. You want this second opinion because management may not understand the real nuances of the use case or task. Only a qualified data engineer or data architect can really give a binding answer.

As this technical person weighs in, they will need to determine whether the task or data in question will be part of a larger data product. Being part of a larger data product could mean that although a piece of data doesn't expressly qualify as big data, it may feed into an overall data product that is big data. This would qualify the data to be a data product exposed by the data teams.

Decide on an Execution Strategy

During the presteps, management should have a rough idea of the project and its initial scope. How long will it take to accomplish? How many people will it take to finish the project? Are these people already with the organization? If they are, how difficult will it be to have them move to the new data team? Will you need to hire from outside the organization to staff the team? How long does it usually take for your organization to hire someone new? Are the people—both internal and external—even competent to execute on the project?

People who are brand new to data and distributed systems will need time to get up to speed. How long will it take for the team to do this? What resources will they need from management? How long should the management team wait after giving enough time and resources to the team to start to expect productivity from it?

The team will need some kind of project management framework to keep things moving and to keep track of their current status. There is a trade-off between providing the structure a team needs to show progress and imposing too heavyweight a project management framework (see Chapter 3). If there's too much process, the team will spend more time going through the process instead of actually doing the work.

Between Planning and Starting

There is a witching hour between planning the project and actually starting it. What happens during this time, and when, will depend on the organization and its data teams. Sometimes these steps happen concurrently during the planning and project start. Other times, they will occur during the planning phase. For newer teams, these witching hour steps will occur during the project start.

It's essential to implement these steps. Omitting them will hold the data teams up.

Find Some Low-Hanging Fruit

Chapter 7 introduced the concept of a data team's velocity. This velocity is how fast the data team can actually work based on the team's growing experience with distributed systems. At the earliest points in the data team's journey, their velocity will be lower. It's at this phase of the project where the data team's velocity meets the level of complexity that's required to create business value.

At this point in the project, you probably have a backlog of things that need to be done for the project. What should you prioritize? What is even within the ability of the team to accomplish at this point? What is the most challenging task in the backlog? What is most crucial to the business? These are all difficult questions to answer and will require engaging with several different stakeholders.

Sometimes teams will aim first at tasks that create the highest business value. Other times, they start with the most challenging part of the project. I suggest finding some low-hanging fruit to start with. This part of the project should show business value but lies within the technical reach of the team. By doing this, management starts to confirm that the data team can do the work, while not taking forever to show some value.

Let the rest of the management team know that you're starting with low-hanging fruit. Some less technically sophisticated managers may think that after this, the data teams are done and that further effort isn't necessary. Instead, the data teams need to quite clear about what is and isn't being done yet. Also, management will need to be clear about how much this effort or task proves the data team's abilities.

Waiting for Data

In any organization, there is data spread around all over the place. Part of the data team's value is to bring it all together in one place to be easily accessible. However, they depend for success here on the rest of the organization, with all its organization diversity and politics.

In this gap, a data team can find itself waiting to get the data for a while. At large organizations, I've seen large teams waiting months for data. At these highly political companies, you're waiting on another team with little or no investment in your success to get data flowing. These wait states are a real killer for productivity.

Here are some examples of wait states I've seen:

- Waiting for the DBA (database administrator) or data warehouse team to bless the data engineering team with credentials to a database so the data can flow into a distributed system

- Waiting for the InfoSec (information security) staff to bless your data movement, method of storage, and security review of the distributed system

- Having to convince another team to start sharing their data with the data engineering team

- Getting a political enemy or someone who feels their team will be affected negatively by the emergence of data teams to release data that they hold onto as a ploy to keep the team from making progress

Management should take care to start the process of getting data as soon as possible. Until then, data teams will be spinning their wheels and never get anywhere because they don't have the data.

Deeply Understand the Use Cases

Earlier steps should have identified the organization's general business needs. Usually, this understanding is done at a high level. At this point in the process, the understanding of business needs to become much more profound, fleshing out the use cases thoroughly.

Chapter 8 talked about what it means to deeply understand the use cases. I've seen teams misunderstand the level to which they need to understand the use case. This isn't just knowing that a team needs to access some data. Instead, the team needs to know when, where, and why they're accessing the data.

It's only after profoundly investigating the use case that the data engineering team should be looking at technology choices. Making any choices on technologies or implementations before this point is fraught with danger. Deeply understanding the use cases lets the team recognize caveats about technologies and choose the right tool for the job.

Starting the Project

After the management and data team has done all of the prework, the data team should have all of the information and data necessary to begin the project.

Getting Outside Help

At this point in the project, management should have a very good understanding of the project, the capabilities of the data team, and the level of difficulty for implementing the project. This should give the management team a good idea of where it needs help from a consultant or a third-party vendor.

The decision of whether to get help requires a great deal of honesty and courage. It requires taking an honest look at the team to see whether they can handle the project's complexity. There is nothing wrong with saying a team needs help, especially if the team is new to distributed systems. The sheer complexity of distributed systems means that most beginners to distributed systems will need help.

Problems in the teams won't solve themselves. It takes a concerted effort to effect the change, and it may take outside help. Some issues may be difficult to self-diagnose. Once again, outside help may be needed to find and fix the problems.

Two common forms of outside help are training and consulting.

Training

Training is a way for individual contributors and management to quickly learn a new skill. Training is especially helpful for new and complex technology. There are a plethora of training methods that have made it easier for teams to learn new technologies. Not every training method is created equal, and there is a trade-off between price and learning.

A good in-person, two-to-three-day training can save a team from two weeks to two months in development time. If the class is taught by a real expert, the team can really understand the technology and how to apply it in the project. If not, the person may be standing in the front of the room reading slides. The tip-off about the quality of a training class can be seen in the price of the course, where a low price on the class means that the budget for the instructor is meager. A low-paid instructor is rarely an expert on the technologies.

Online and virtual classes have the highest degree of variability in quality. When looking at online training materials, the management team should keep in mind the final goal for the training. If the management team wants the data team to simply have a cursory knowledge of the technology, an introductory-level class will be sufficient. If the management expects the team to actually implement the technology in a project, an introductory-level class won't be enough. The majority of online and virtual courses may market themselves as "master level" or "in-depth," but one can quickly see that the class is too short to deliver mastery of the topic.

In some ways, the training industry is its own worst enemy. Once a technology becomes hot, the training industry will churn out products of questionable quality that are focused on price and not real success. When looking at a training course, remember that a cheap material will be more expensive in the long run than a more expensive class taught by an expert. Watch out for companies putting forward warm bodies to read slides.

When done right, training classes will have the highest ROI for the data teams. The team is taught how to create their own solutions instead of having the solution designed for them. Instead of learning through trial and error, the team can accelerate their learning with an expert.

Consulting

Consulting is another way data teams can get help. Different levels of consulting engagements range from high level to deep in the weeds. At the deep end, consultants write all of the code and create the entire solution. Consulting is useful to accelerate the team, and management should decide on what is the right level, given the team's abilities and ambitions.

Depending on the knowledge of the consultant and the complexity of the project, the consultant can accelerate the team. However, the consultant will be doing the majority—if not all—of the intricate pieces in the project. The data team won't be increasing its velocity. This could be a red herring to the rest of the organization, where it appears the data team is making progress and doesn't need the consultant anymore. The reality can be that the consultant was doing all of the work, and the team themselves have not deepened their expertise. Depending on the level of the team's expertise on the technologies, the organization can allocate extra time and budget for knowledge transfer from the consultancy to the organization.

When choosing a company for consulting, watch out for consulting companies that have just started out with distributed systems, programming, or data science. This can lead to the helpless (your organization) being led by the clueless (the consulting company). In this situation, your organization may not have the experience or knowledge to know that the consulting company is clueless. Organizations can mitigate some risk of choosing the wrong consultancy by asking for recommendations from people with direct knowledge of a consulting engagement that went smoothly.

I've been in the situation of having to tell a client that their other consulting company is clueless. To be clear, this wasn't a difference in opinions. This was a situation where the consulting company was missing the programming and distributed systems skills entirely. They had learned enough to say the right words and make a little bit of progress. Once the project was slightly more complicated than the consulting company's minimal understanding of distributed systems, they couldn't make any more progress.

Choose the Right Technologies

The choice of right technologies comes only from a deep understanding of the use cases. Technology choices can play a significant role in the success or failure of your project. Choosing the wrong tool for the job leads to collapse, and selecting the right tool leads to success. A significant difficulty for new teams is sifting through the sheer number of distributed systems technologies that are available. Although the data teams will need to implement and use 10–30 different technologies, there is a pool of 100+ other technologies that they will have to sort through.

The process of sorting through all of the technologies can take a significant amount of time if the team has no exposure to distributed systems. This means that the team is held up and waiting for architecture and technology decisions. To avoid rewriting code later, the team must spend some time researching the right strategy, and those are expensive wait cycles.

As new teams start to choose technologies, they will often gravitate to highly touted technologies that others in the industry consider to be the latest and greatest. There is an inherent risk using the latest and greatest technologies. For example, a team could choose a technology that is new but lacks the community around it to get support and figure out issues. Often, the technologies that have been around for a while are a better initial choice. They may not be as cool and trendy, but they will have better odds of success.

To make the right choice, the team needs a qualified data engineer or project veteran in the room. This expert brings their experience in putting distributed systems into production and in the issues that go along with long-term usage of distributed systems. This experience in choosing technologies will help mitigate some risk of selecting the wrong technology or the wrong technology for the long term.

As a data team starts to choose their technologies, they may gravitate toward distributed systems, but not every use case requires a distributed system. Always look at the scope of the use case and the potential growth in data or processing before choosing a distributed system. Using an unnecessarily complex tool for the job, in this case, could mean a massive over-engineering of a use case. The team will spend far too much time on something that should have been completed much faster. "Faster," in this context, could indicate both the amount of time it took to write the code and the amount of time a query takes.

Other times, the initial code—whether small data or with a distributed system—was written so poorly that it is the real source of problems. Although changing systems may help, it's more important in this case to rewrite the poorly performing code.

Choosing technologies is an iterative process that works in conjunction with your crawl, walk, run strategy, and the data team's increase in velocity. As the team progresses, they will need to add new technologies required for the next step. The initial technology choices need to be in line with what the team can handle at their current velocity. As the team's velocity improves, new technologies can be added to better manage a use case or even begin an advanced use case.

Write the Code

It's only after much prework that the team can start writing the actual code. This level and amount of prework stands in contrast to small data systems. Smaller projects never required this level of prework before writing the first line of code. Because of their background in small data, some teams will just start coding without doing their prework. This is another source of failure for teams because the technologies may need to change, and there isn't a deep enough understanding of the use cases to code correctly.

Depending on the phase, the quality of code could vary. The team could be writing prototype-level code. It's important to know that a beginner team's prototype code will need to be thrown away or completely rewritten. Because of this, using prototype code in production leads to technical debt that the team will have to pay down.

It's at this point that the management may want to revisit their decision to get help. If management chooses not to get help, it should closely monitor the data teams to see whether they're getting stuck or whether progress is glacial. In these cases, the management team may want to get help immediately. If management did choose to get help, the value that the outside consultant is providing should be apparent. If the consultant is providing little to no value, management may need to make some personnel changes.

Create the Compute Cluster

The data teams will need to create several compute clusters. These clusters will enable a team to run or test their code at scale. The amount and architecture of these clusters will depend on the team or individual requirements. For example, what type of cluster resources will the data engineers develop and test their code? What type of cluster resources will the data scientists need to develop and test their code?

The data engineering and data science teams may need both individual virtual machines and clusters. The individual virtual machines will be used to develop the code while running the distributed system's processes on a single-machine cluster. The clusters will be used to test code and performance at scale. This division allows for easier debugging and faster code writing.

The quality assurance (QA) team has their own requirements. How will QA run their tests in both an automated and manual way? How will QA do load testing or performance testing? The QA team may need several clusters or several types of clusters. If the organization is deploying on-premises, the QA team's hardware should match what is in production exactly. Otherwise, the load testing and performance testing results will not match what the production cluster can do or how it is set up.

For many organizations, using the cloud is a great way to spin up data engineering, data science, and quality assurance clusters. Using the cloud gives great flexibility in how clusters get spun up and used. Many organizations automate their cluster spin-up. This allows an individual to stop their cluster as soon as they are done with their task. This gives organizations the best of both worlds, spinning up each cluster only as long as it is needed and stopping it to save money.

Create Metrics for Success

Often data teams will have clear use cases of what needs to be accomplished. They will have a clear backlog of what needs to be done and for whom. However, they lack a clear metric for what success looks like. Without a clear metric for success, they may feel like they can't succeed or that the goalposts are continually moving. In either case, a lack of metrics for success will make the team feel as if they've never accomplished a goal.

Goals and metrics will be different for each team, project, and organization. These goals are usually based on an overall organization goal and team-specific contributions to that goal. They're commonly tied to a business objective or to enabling the use case for the business. Other metrics are related to interactions with other data teams. Here are some examples:

- Improving how the data engineering team exposes data products

- Reducing the amount of time data scientists spend on engineering tasks through better data products

- Reducing operational downtime through better metrics and alerting

- Improving efficiency and optimizing the machine learning models created by the data science teams

Iterative Steps

I highly recommend an iterative approach to projects. Some steps should be taken over more extended periods and each phase of the crawl, walk, run.

Evaluate

Based on your metrics, goals, and use cases, you need to evaluate the project and data teams.

The critical question is, are you successful? Success can mean many different things. Did you gain value from the data in your organization? Is the team increasing velocity? Did you create real business value from the data products, or are they sitting there waiting to be used?

As you look at the goals, did you beat or fall short of your goals? More than likely, your organization will have a mix of goals that were achieved and other goals where the team fell short. Why did you fall short or beat the goals? Does the team need more help? Does management need to make some changes to the team?

To really get to the root of problems and get a 360-degree view of the issues, I highly recommend doing a blameless postmortem. During this meeting, the team talks about what went right and what went wrong in a blameless way. It's crucial to have all voices in this meeting to get a holistic view of the issues. For example, if the postmortem is done without the quality assurance team, the problems of how the data engineers are passing off code to be tested aren't heard. The team may not understand that the unit test coverage is far too low and happy-path focused.

If these meetings become full of blame and more focused on pointing fingers, the value of the meetings will go down. People on the team will become defensive and shut down, while the loudest voices are the only ones heard. During these meetings, if a person hasn't spoken, I will specifically ask them if they have feedback or something they want to mention. In any case, it's crucial that the manager leading these meetings be a moderator to keep the meeting running smoothly.

When to Repeat

Many, if not all, of these steps need to be revisited for every phase of a crawl, walk, run sequence. The amount of time between these phases can be months or a year. During these periods, things should be changing. The team should be getting more experience. The business side should be changing, and its use and expectations for data products should be increasing.

For example, what would be the difference if the presteps weren't redone between the crawl and walk phases? The data and data problems should have changed. At least crawl-level data should be in place. The business is now looking at data in a previously impossible way (the original "can't"). The business side should be seeing new things that they want to do with the data.

Also, the data team's velocity was at a crawl level. All kinds of business desires were outside the velocity of the team at that time. Is the team ready for the next jump in complexity?

It's for reasons like these that I recommend an iterative approach to work on projects. An organization is always in flux, and the data teams need to be iteratively reacting to these changes.

CHAPTER 12

Organizational Changes

You ain't seen the best of me yet
Give me time

—"Fame" by Irene Cara

Creating data teams isn't just a matter of acquiring some new skills and deploying a cluster. Rather, it takes some organizational changes to do it right. Some of these organizational changes can be painful to implement. This chapter covers some of the main organizational changes that will have to happen alongside technical ones.

- How to work with and gain the trust of the old guard

- What should be done with any previous teams

- The innovation and progression that should be coming out of the data engineering and data science teams

- How to assess and deal with skill gaps on the teams

- The hardware changes that teams need

Working with the Old Guard

At large organizations or organizations that have been around for a while, there will be a group of people who will reflexively resist any organizational change. This old guard isn't about the age of people, but about people being comfortable in their job. Changes with data, data teams, and new technologies will put the old guard out of their comfort zone. The old guard can bring any new project or transformation to a screeching halt through their inaction or working against the change. It's up to management to gain the buy-in of the old guard.

© Jesse Anderson 2020
J. Anderson, *Data Teams*, https://doi.org/10.1007/978-1-4842-6228-3_12

Getting buy-in for change can be a challenging proposition—especially one with such extensive organizational and technical change as creating data teams. Management will want to explore the reasons why a move would be troubling to the old guard. Are they worried about losing their jobs? Are they concerned about if or how they can learn the new technologies they'll be expected to learn? They could feel as though the old system is their baby, and they're worried about what the new system is going to do to their baby.

The management team really needs to get ahead of these questions with incisive answers. Otherwise, the rumor mill will start churning, and people will get all worked up over a lack of information. I've dealt with companies who put their old guard into a vacuum of information, and it took us much longer to back out all of their fear and worries. It's much better for management to provide answers as early in the journey as possible.

Getting the old guard's help when creating new data pipelines is critical on a technical level. Software and organizations that have been around for a while have years and years of stories, decisions, and architectures that the new data pipelines will need to integrate with, receive the data from, or adapt to with all their quirks and nuances. It will be your old guard that has the knowledge of these issues and will be at the forefront of getting the entire system to work correctly. Management should internalize this fact and be working with the old guard to create as frictionless a working relationship as possible.

I've also experienced the times where—despite our best efforts—the old guard was not going to help us. The new approach to data scared them too much. They made it clear they were going to work against us and make any integration as time-consuming as possible. This is where the problems become both organizational and technical. The management team may need to start applying pressure on the old guard's management. Most of the time, the pushback really doesn't work because the person has tenure, and the old guard's management is more worried about rocking the boat. The technical team may have to start reverse engineering the systems to see how they work and make educated guesses about how to integrate. These situations take more time than an average project, and management should adjust the timelines and expectations accordingly.

What to Do with Previous Teams?

Not every person will be able to make the jump to a data team. Not every nondata team that was creating or working with data products will need to stay at the same headcount. That leads to what is honestly one of the most challenging sections to write for this book. It potentially means peoples' jobs and livelihood.

In a basic sense, the data teams will supplant some use cases or will significantly reduce the need for some other teams. In some cases, data teams will altogether remove the need for an entire nondata team.

A reduction in headcount for nondata teams will put a new emphasis on workforce planning. Often, there is a small technology serving a data product in production usage, and the management team has the plan to completely replace the data product with distributed systems. This new data product will be created and operated by the data teams. Over time, the small data team will have much fewer responsibilities, and the data team's responsibilities will significantly increase.

What about moving old staff from small data projects to distributed systems? This is usually difficult for both technical and nontechnical reasons.

For some organizations, teams, and people, their current workflow is something they've spent years creating and perfecting. They're quite comfortable in that space, and changes sound difficult and problematic. These staff may not sense the limitations their approach is placing on the organization as new data needs arise. This can even extend into work practices, where there is a clear separation of concerns between development and operations. As a data team moves to DevOps, the operations team is now forced to upskill in programming skills, and the data engineers are forced to upskill in operations.

For others, the silo around their data is perceived as a moat for job security. If another team skirts their moat or makes the data far more accessible, that triggers a defense mechanism as they fear that their moat is failing.

Some staff have made a long-term career bet on a particular technology. For many years they never reassessed their long-term bets and doubled down on their previous strategy. There could be people with 10–20 years of experience in data warehousing, mainframes, or another waning technology choice that will feel threatened by the data pipelines and projects. These changes will introduce new workflows and organizational changes that fly in the face of their comfortable and long-term bets.

Innovation and Progression in Data Engineering and Data Science

Data engineering should continue iterating by making better and better data products. This includes creating higher quality data for the data scientists and for other analytics purposes. It also includes adding new technologies to improve the speed and scale of data products.

Data engineering also implements new use cases. This is part of the growth of demand within the organization for data products. The data teams need to be keeping an eye out in the organization for the emergence of new and required data products. The data engineering teams will be the ones to create them.

Data scientists should also be engaged in continuous improvement by upgrading their models. This improvement can come from better quality data or from tweaking various parameters. In both cases, the models will need to be retrained to gain the maximum value.

New research is continuously coming out on how to improve an existing model or a brand-new model that has higher accuracy, is faster, and so on. Data scientists will need to have time and space to read the latest research. This may include some time to try out the new approach and see whether it improves upon the current approach.

ARE DATA TEAMS PERMANENT?

Often, management wonders whether a data team is a permanent or temporary addition to the organization. The simple answer is they are permanent. Organizations that don't realize this haven't come to a full appreciation of data culture. This could be due to a lack of education on data teams and what they do, or to another, more problematic issue.

Data products need to be treated like any other product the organization develops or makes. Data products can't be short-lived projects that are delivered and then be done entirely. This concept may be new to organizations because continuous improvement and additional data products have to be made for data scientists.

Data scientists' ability to create accurate models is directly dependent on the feature quality of the data. This data will change over time. Keeping data quality high has to be a continuous investment, so data teams must be permanent additions to the organization.

Management—especially the management of the data teams—needs to educate the rest of the organization about what the team does and the value they provide. Likewise, the onus is on the management to actually create that value for the rest of the organization. Data teams that aren't providing value create a huge problem that needs to be addressed.

The value being provided from the data teams should be growing. Honestly, if the use cases, demand for data products, and usage aren't growing, you're doing something wrong. Organizations grow their data teams organically because the value is so high, and the demand is always outstripping supply. If there is too much supply and very little demand, the management team needs to figure out what is happening.

Assessing Skills Gaps

Part 2 talked about the skills needed for data science, data engineer, and operations teams. The number of people employed and their skills will vary depending on the complexity of the project, the current velocity, and the budget for headcount. However, each of the skills must be possessed by at least one person on the team. The complete absence of a skill from a team should raise a red flag.

The job of looking for missing skills is called a *skills gap assessment*. This should be done by management or a lead on the team.

A skills gap assessment requires the utmost honesty with yourself and the person being assessed. I recommend setting aside some deep thinking time to really go through this skills gap assessment.

There may be skills where a manager doesn't entirely know if the person possesses the actual skill. This could require a follow-up with the person, or the manager could ask the technical team lead for their assessment.

Assessing Levels of Expertise

Some gaps won't take the form of a skill. Instead, these assessments will be on the experience levels of the team. Some teams have more stringent requirements on experience levels. For example, Chapter 12 covered how a data engineering team's levels of experience are beginners, qualified data engineers, and veterans.

When performing this gap analysis, you will need to give the people a rating on their experience level. As we finish the gap analysis, we'll check to make sure we have the right experience levels.

How to Perform a Gap Analysis

Take a piece of paper and turn it widthwise, or open a spreadsheet and create a column for every skill the team should have. Then create rows for every person on the team. On the top row, write all of the skills that a data team needs. Next, list the names of everyone on the team in the far-left column. See Figure 12-1 for an example of how to do this.

	A	B	C	D	E	F	G	H	I
1		Distributed systems	Programming	Analysis	Visual Comm.	Verbal Comm.	Project Veteran	Schema	Domain knowledge
2	Mohamed								
3	Gabrielle								
4	Mateo								
5	Joseph								
6	Feng								
7	Adam								
8	Gabriel								
9	Jack								
10	Riya								
11	Juan								
12	**Total**								

Figure 12-1. *A screenshot of a spreadsheet showing how to lay out the skills and people*

Now you're going to fill out the paper or spreadsheet software. Go through everyone on the team and put an "X" whenever that team member has that particular skill, as shown in Figure 12-2. This requires a frank look. Doing this analysis as honestly as possible could be the difference between success and failure for the team.

	Distributed systems	Programming	Analysis	Visual Comm.	Verbal Comm.	Project Veteran	Schema	Domain knowledge
Mohamed	X	X						
Gabrielle		X						X
Mateo		X		X				
Joseph		X	X					
Feng		X	X					X
Adam		X						X
Gabriel		X			X			
Jack		X			X			
Riya	X	X				X		X
Juan		X						X
Total	2	10	2	1	2	1	0	5

Figure 12-2. A screenshot where the person has been marked as having the skill

Repeat this gap analysis for all three data teams.

Sometimes, managers will want to give the person an emerging status or half a point for a skill. Sometimes people will do their gap analysis and put percentages or numbers with decimal points instead of an X. They'll put a 0.1 or 0.5 instead of X. This will just keep you from facing a real skills gap. The skill is either there, or it isn't. In my experience, ten people with a "0.1" are no better than giving the team a zero. The skills gap assessments are very binary because the individuals have to meet a certain threshold.

Interpreting Your Gap Analysis

Now that you've placed the Xs on everyone's skills, total them up. This will tell you a story about the makeup of your team. It's going to tell you where you're strong and weak, and maybe whether your team can't even get their work done. If there is a separate manager for each team, the data team's managers may need to get together to compare their notes on where there are gaps.

As you look at the gaps, start asking yourself some key questions. Was the assessment of skills done accurately? How critical is the missing skill to the success of the project? Is the missing skill necessary now, shortly, or further out? How difficult will it be to train someone to get this skill, and is the need urgent enough to hire someone new? How long will it take to hire someone to fill in this gap, and will that put the project behind?

Some teams need certain skills more than others. For example, on a data engineering team, the majority of the team needs to have distributed systems and programming skills.[1] If they lack one of the necessary skills entirely or have very few people with the skills, the project may take way too long or may not even be possible. The lack of programming skills and distributed systems skills are often associated with teams that drew their data engineering team from their data warehouse team or another SQL-focused team.

It may be difficult or impossible to hire people to fill some skills gaps. In organizations with substantial domain knowledge requirements, it may be easier to take existing staff with the domain knowledge and train them on the new skill.

You also need to look at the experience levels of the team and some pressing questions. Is everyone on the team a beginner? Do I have at least one qualified data engineer? Do I have access to or have a team member with veteran experience? Does the team's experience level match the required level for the complexity of use cases? Higher levels of experience can't be passed on through training but require firsthand experience. That's why Chapter 4 described the crucial role of a project veteran, who must come up through the ranks.

Ability Gaps

Through most of this section, we've been assuming that each person on the team can acquire the skills you need. A skills gap means that a person currently lacks the knowledge, but has the innate ability to learn the skill. This isn't always the case. There could be an ability gap, which means that a person lacks both the knowledge and innate ability to successfully learn the skill. No amount of time, resources, or expert help will get them to that next level.

In data teams, there may be several examples of ability gaps. For example, visual communication may require an innate artistic ability that can't really be taught. Expecting a person without a creative flair to create something that looks and acts beautifully may be unfeasible. Likewise, not everyone can understand distributed systems or programming at the level that's required for data teams. Expecting someone who is brand new or has never programmed to create a distributed algorithm may be virtually unfeasible.

---

[1]For more information on interpreting a gap analysis for data engineering teams, see my *Data Engineering Teams* book in chapter 3, "Data Engineering Teams."

When doing a skills gap assessment, the managers need to think about whether the gap is one of skill or ability. If the person can't eventually acquire the skill, the management team will need to make other arrangements. This could include hiring another person to complement them or finding another place in the organization where their talents are a better match.

Hardware Changes

Some organizational changes revolve around hardware. Distributed systems lead to new hardware requirements. These hardware requirements can differ in speed, size, and price from hardware the organization used before with small data.

Cloud Usage

If at all possible, I recommend using the cloud. This way, hardware acquisition is removed as a blocker before the POC (proof of concept) can start. Some popular distributed systems technologies are already available as a managed service. These managed services eliminate the need for the team to learn how to install and start the distributed system. Instead, they can start programming with the system.

The cloud will allow the team to spin up and spin down resources as needed. This allows the team to control their costs on a fine-grained level as they develop and test the system. If the team somehow completely botches or destroys the system, they can easily spin the cluster back up. The team can control prices for the cluster by shutting it down when it isn't being used, such as overnight or during the weekend.

Purchasing Clusters

If an organization chooses to place its clusters on-premises, the organization will need to purchase the hardware. This hardware can be more expensive than small data servers because of higher redundancy requirements. Higher redundancy can include redundant power supplies, but the primary differences lie in storage and networking.

In distributed systems, data is stored multiple times—ranging from 1.5 to 3 times—in its entirety on various computers. All of this redundant data is moved over the network from computer to computer. When data is being read for processing, this data can flow over the network again.

Some organizations expect or have mandated that all on-premises infrastructure use virtualization. There are apparent trade-offs between virtualization and bare metal,[2] mostly around ease of operations. In my opinion, the big reason to do on-premises vs. cloud is to have the choice between deploying with virtualization or on bare metal. Some distributed systems may take advantage of the increased speed and lower latency of bare metal deployments.

Systems for Big Data Teams

Hardware changes and upgrades aren't limited just to clusters. An often-forgotten portion of an organization's big data efforts is located in hardware and data requirements for the people writing the code. These hardware requirements primarily affect the data engineering and data science teams.

The main reason behind these changes is that big data is actually big. The individuals may need to store large amounts of data while they are developing code. They will also need to run the distributed systems on their own hardware. Few of these distributed systems are optimized to run on computers with low memory or CPU. All of the distributed system developer's effort goes into running efficiently on a cluster and not on a developer's computer. These heavyweight distributed systems will need more CPU and memory than a small data developer. That means that standard-issue hardware may not be enough for developers.

Development Hardware

It's essential to consider the hardware that the data engineering and data science teams are going to use. During development, it's often better to have the distributed system running locally on the person's computer. This removes many variables of what can go wrong or be changed if developers are using a shared resource for their distributed system. If they already have hardware such as a laptop, it may not have the power to run the distributed systems that they're developing code for. Management may need to start early either to buy new computers or to upgrade their existing computers for the new hardware requirements.

---

[2]Bare metal refers to a single-tenant setup where all software is installed directly on the system itself. Virtualization refers to a multitenant setup where resources are allocated and isolated using a virtual machine. Some cloud vendors have seen the need for single-tenant systems and support these configurations.

Some organizations put their developers on virtual machines running on a separate computer for development purposes. These virtual machines can suffer from the same issues of not having enough resources. Management may need to start lobbying for a new level or class of virtual machines for development with distributed systems.

Some organizations lock their computers down and prevent their developers from having root or administrator access to the operating system. This can really slow down development because developers need to install and run the distributed systems on their computers. Without root or administrator access, they won't be able to. Some organizations have an approved software list that makes it easy to install the software. More than likely, these distributed systems won't appear on that list. Management will need to figure out how to get root access or work with IT to get the new software installed.

Test and QA Hardware

Organizations often forget about the hardware needed for quality assurance (QA). They need more than a single cluster. QA may need several different clusters for each phase of their testing. They will need one cluster for validating the functionality of the software and another for performance and load testing. The hardware for performance and load testing should match exactly what is in production, but testing for validation doesn't necessarily need to be the same.

If the QA team lacks the hardware resources they need, this could harm the quality of the data products. A lack of hardware could delay the release of new software because the QA team is waiting for the hardware, or because their hardware isn't fast enough. If the QA team doesn't do performance and load testing, the new software could have a performance bug and not perform as fast as the previous release. Performance issues in production can be stressful and challenging to find. It's much better to find and fix these issues through performance and load tests before releasing the data products.

Getting Data to Data Scientists

To get their work done, the data science team needs access to data representative of the data used in production. The management, data science, and data engineering teams will need to figure out the best way to get data into the hands of the data scientists. Without the best data possible, the data scientists will see a skewed representation or not encounter all of the edge cases.

Getting data into the hands of data scientists can be difficult for several reasons. A common one is the security implications. What should an organization do if the data has PII (personally identifiable information) or financial information? Who should have the responsibility to choose the best way to deal with PII? Who should be the ones to implement the methods of dealing with PII?

Another issue may be the sheer size of the data. How can the data scientist do discovery or analysis on 1 PB of data? The technical abilities of the data scientists need to be kept in mind. Data scientists may not have the engineering knowledge to choose the right tool for the job. This is when the data engineers must help the data scientist understand what they need and how to achieve it with the right tools.

MANAGING RISK IN MACHINE LEARNING

As machine learning (ML) gets embedded into more systems and applications, companies are becoming aware of specific challenges that come with products that combine data, models, and code. The umbrella term I've come to use is risk, and over the past couple of years, I've been giving talks on how companies can go about managing risk in machine learning.[3]

The advantage of framing the situation in terms of risks is that certain industries (such as finance) already have well-defined roles and procedures governing risk management. ML is still relatively new within many companies. Because it combines data, models, and code, machine learning systems pose challenges that need to be accounted for from the initial planning stages, all the way to deployment and observability. These challenges and risks include:

- Privacy
- Security
- Fairness
- Explainability and transparency
- Safety and reliability

---

[3]www.oreilly.com/radar/managing-risk-in-machine-learning/

Risk management in ML will require teams that incorporate people from different backgrounds and disciplines. Having teams composed of people from diverse backgrounds ensures a broader perspective, which in turn can help prevent the release of models that exhibit gender, cultural, or racial bias. Depending on the application, experts in privacy, security, and compliance may need to be involved much earlier in the project. If machine learning is going to eat software, we will need to grapple with AI and ML security too! This will mean incorporating experts in security and privacy. It's extremely rare to find data scientists or machine learning engineers who are well versed in the latest developments in cryptography and privacy-preserving ML, or who know the latest defenses against adversarial attacks on ML models.

Moving forward, we'll need to have better processes and tools in place to help teams manage risks in ML. In a world where ML will be increasingly prevalent, we also need to rethink how we organize and staff product teams. Companies will need data teams with more diversity in terms of background and skill sets.

—Ben Lorica, Author, Advisor, Conference Organizer, Principal at Gradient Flow Research

CHAPTER 13

Diagnosing and Fixing Problems

Work it harder
Make it better

—"Harder Better Faster Stronger" by Daft Punk

This chapter covers some common problems that I've seen as I've worked with data teams all over the world. The solutions to the issues will vary. As a data team manager, you will need to figure out the true cause of your problem. This is the most challenging part of data team management. Your diagnosis will determine the success or failure of the project, and perhaps the data team as a whole.

If you skipped all of the previous chapters to get here and figure out your team's problems, I highly recommend that you go back and read the rest of the book. As I give possible fixes, I'm expecting you to understand and have read all of the previous concepts.

Some issues will seem to be similar to another issue—to the untrained eye. To the trained eye, the issues are different because the root causes are different. I've sought to break out these issues because each root cause and solution is unique and needs to be fixed a different way.

Stuck Teams

Sometimes teams just can't make progress. Every daily stand-up sounds the same. Every status report looks like the one from months ago. The team may or may not realize it is stuck, and management will need to keep an eye on a lack of progress. These teams and problems don't solve themselves. The teams may think they can pull it off, but my experience is that they don't. These problems usually stem from technical or personnel issues that are too tough or entrenched for the team to solve on their own.

© Jesse Anderson 2020
J. Anderson, *Data Teams*, https://doi.org/10.1007/978-1-4842-6228-3_13

The team says they'll get things done, but they said the same thing a month ago

When the team's status sounds like the movie *Groundhog Day*, the team is stuck. In *Groundhog Day*, the main character wakes up to the same day and repeats the same day over and over. Similarly, a stuck data team will wake up every day and do the same things as they did the day before. There will never be any real or substantial progress.

There are a few common root causes to stuck teams:

- The team doesn't understand what they're doing. The complexity is too high, or the team lacks the necessary velocity to even attempt the project.

- The team doesn't feel comfortable telling you there's a problem. This can be a cultural issue that management should be aware of.

- You are using consultants who don't know what they're doing. Because the data team lacks the knowledge of distributed systems or programming, they may not be able to know when something is wrong. They can see only the outward manifestation that the consultants can't or don't make any progress.

Whenever I give the team a task they come back and say it isn't possible

In a healthy organization, a data team creates data products that get put to use by the business. As the organization becomes more and more enabled by data, their asks and requests should become more and more complex. But at some point, the data team may push back on a request by saying that the task isn't possible.

In this circumstance, there is a difference between impossible and time-consuming or tricky to implement. With impossible, the team is saying that there is no technical path to accomplishing the task. That may or may not be the case. When the team says the task is time-consuming, they are pushing back on the relative priorities they have before them. They are asking, "Do you want us to do this task or another one that will take less time?" Management will want to be careful to understand the priority and technical difficulty of each request.

When a team is saying no consistently, this usually comes from one of the following:

- There is a lack of experience on the team. This could be a lack of experience in the domain or with the technology itself. The team may not have the velocity or foundation to accomplish the task (see Chapter 7).

- There are only beginners on the team. To a beginner, everything will seem complicated and impossible to do. With the right mix of experience on the team, more tasks will be possible (see Chapter 10).

- There's no project veteran on the team. Without the guidance of a project veteran, the team may be unable to figure out a way to accomplish the more complex tasks and simply say it's impossible to save face (see Chapter 4).

- The team doesn't have the right people. For example, the data engineering team could have data engineers who don't meet the definition of data engineers. This what are ostensibly data engineers that are unable to accomplish the task.

I can't tell the difference between the team making progress and being stuck

For some managers, it's difficult to gauge what progress—if any—the team is making on the project. This mostly happens when the management team lacks technical knowledge or hasn't led data teams before. In this case, the management team will have to figure out a way to get an accurate assessment from the team on their progress.

In some cases, increases in project complexity raise it past the abilities of the team members. In these situations, the team themselves may not even realize that the project is outside of their abilities until it is too late.

Getting an accurate assessment from the team requires that the management team has established a good rapport with the team. If there are already problems with the team or project, the trust between the managers and individual contributors could be in jeopardy. It's essential to have already built a trusting relationship and have a good rapport before the problems start.

Once there is a good rapport between management and individual contributors, the team will be more comfortable about sharing their actual status on the project.

In some cultures, there is a deep reticence about admitting failure or acknowledging an impending failure. Management should understand any potential cultural differences and keep an eye out for cultural issues that prevent the actual status from being reported.

We have trouble with small data systems, and we're having even more trouble with big data

The significant increase in complexity between small data and big data can be manifested in several different ways. Organizations that assume that big data systems are just as easy as small data systems will fail. If an organization can barely make it with small data systems, they'll have extreme difficulties with creating big data systems and the distributed systems that the team will need to use.

In situations like these, the organization will need extensive outside help. Without it, the organization will likely never achieve its big data goals. Depending on the levels of the team's abilities, training may be an option but will have a limited utility. If the team is barely able to understand small data systems, they won't understand or be able to act on instruction about the much more complex distributed systems. More than likely, the organization will need extensive outside consulting to run the entire project, while the members of the organization provide domain knowledge and program management.

Underperforming Teams

Underperforming means that the organization is not gaining the maximum amount of value from data. Often, these teams or organizations don't know or realize they're underperforming. The team ostensibly ticks all of the boxes for data products, but when pressed to improve the data product or create a more complex data product, the team can't do it.

There is a difference between a team that is stuck and one that is underperforming. When a team is stuck, it doesn't make any appreciable or real progress on the project. When the team is underperforming, the team does make progress, but the growth is limited in its scope or complexity.

Whenever I give the data team a task, the team creates something that doesn't work

At the crawl phase, the team was able to make progress and create data products. Now that the project becomes more complex—such as at the walk or run phase—the team can't create something that works at a development level or in production.

When a team is unable to create something that works, this usually comes from one or more of the following:

- There is a lack of qualified data engineers on the team. The code being written lacks the fit and finish that comes from a qualified data engineer. The qualified data engineer will have seen firsthand the need for good unit tests and integration tests that put high-quality code into production (see Chapter 7).

- There are only beginners on the team. When a person is brand new to distributed systems, they will create the worst abominations possible. These abominations don't work in production (see Chapter 10).

- The team is untrained on the technologies they are supposed to be using or have a velocity insufficient for the project's level of complexity. The team may need more training and time to understand distributed systems before continuing on to more complicated projects.

- There is a lack of operational excellence in the organization. The organization may require an operations team that has been trained on the distributed systems they're expected to maintain.

Whenever I give the team a task, they create something that isn't really what I asked for

Here, the business side has asked the management of the data team for a new feature or data product, and the team says that they understand the job. The team works on it and comes back with the data product. When the management or business goes to use the data product, it isn't usable or doesn't do what the business asked for initially.

Unusable data products usually come from one or more of the following:

- The data engineering team lacks domain knowledge. To really create a usable data product, the data engineering team needs to understand the use case in great detail. Without a background in the domain, the team can implement only a smaller amount of the project requirements.

- There may not be enough BizOps or management involvement. Sometimes, the management or the business hands over the tasks to the data team and backs away. There should be an ongoing and consistent interaction between the business and data teams to make sure the data products meet their needs.

- The data team may not be competent. They may not meet the qualifications set out in their job descriptions.

- The instructions that are given are vague. Although the business and management think the instructions are clear, they aren't clear enough, or there are too many nuances that aren't spelled out clearly.

- This is the best that can be done given the various circumstances. Distributed systems are complicated, and there could be a technical reason or limitation, making it impossible to deliver the data product requested by the business or management.

The team can do elementary things, but they can't ever do something more complicated

At the beginning of the project—such as at the crawl phase—or when there is a straightforward task, the team can accomplish the task or create the data product, which might involve writing an SQL query or expanding the use of an existing data product. But whenever a more complex task comes in—such as at the walk or run phase—the team can't write the code or create a more complex system.

This inability to do complex tasks usually comes from one or more of the following:

- The team may have skill gaps. For example, the data engineering team may be completely missing the programming or distributed systems skills. The team may still have these gaps even after providing the learning resources for programming or distributed systems. At this point, the gaps become ability gaps (see Chapter 4).

- The team is made up only of beginners. Such teams can create systems only at a very low complexity (see Chapter 10).

- The team is untrained on the technologies they are supposed to be using. They may have a fundamental misunderstanding of the necessary technologies and can't do something more complicated as a result.

Skills and Ability Gaps

The source of problems on a team could lie in a skill or ability gap. A skill gap means that a skill is within reach of the person, and it is just a matter of time or resources for the person to acquire the skill. An ability gap means that a skill is not within the person's reach. There is no amount of time or resources that will allow the person to acquire the skill.

Chapter 12 explained how to identify skills gaps. Managers can rectify skill gaps through training or allowing staff to gain more experience, whereas ability gaps have to be fixed through personnel changes. The actual personnel changes will depend on the scenario and skill that is missing.

Skill and ability gaps should appear only when moving or migrating an existing team or individual to the data team. If you are experiencing this while creating or hiring a new data team, you aren't hiring the right people. When migrating, the management team may lack the experience or foresight to know which people in the team have a skill or ability gap. The only way to establish the truth is with time, effort, and resources.

How can I tell the difference between a skill gap and ability gap?

It can be challenging to figure out or tell the difference between a skill gap and an ability gap.

Start by looking at what has been provided to the individual. Have you provided them a viable and quality resource to learn the skill? Not everyone can learn from books or videos. Was the person given a chance to learn in-person from an expert? What was the quality level of the resource that was provided for them to learn from? If the learning resource was of low quality or very introductory, it might not have conveyed the necessary skills.

Time is also a resource that management needs to provide. Is the person expected to learn new skills while continuing to perform their current tasks and job? Does the management team expect the person to learn the skills on their own time—spending their nights and weekends on it?

If management hasn't provided enough resources to the person, the problem may be a skill gap that can be rectified by starting to provide the necessary resources and time.

If management has provided enough time and quality resources to learn the skill, then it's time to look for ability gaps. When asked why the person is having difficulty learning or understanding the concepts, what do they say? Do they fault a lack of time? Do they say they need to get around to it? These procrastination and delay tactics are often the marks of an ability gap. Management needs to be resolute in determining that a new person must be hired for the task.

Is the team having trouble learning how to program?

Different teams need different levels of programming ability. The operations team needs relatively modest skills. The data science and data engineering teams, on the other hand, require serious programming abilities—with the data engineering team having the most need.

If a data scientist doesn't know how to program, there may be title inflation because the person may be closer in skills to a business intelligence analyst or data analyst. Data engineers should come from a software engineering background. If a data engineer doesn't know how to program, they didn't come from a software engineering background.

If programming skill is low or completely missing on a data science or data engineering team, the organization should not even attempt a big project until it makes significant fixes to the team. If the lack represents a skills gaps, the team will need to start undergoing very intensive training on programming. If the lack represents an ability gap, the organization will need to start hiring from outside.

The team is coming from a small data background and having difficulty learning big data and distributed systems

Having trouble with distributed systems will vary in gravity, depending on which team is having trouble.

The data science team has the lowest bar with distributed systems. They may need to know only enough to get by with a lot of help from the data engineering team. For instance, they may need to learn how to write distributed code for machine learning or other processing. An ability gap in distributed systems is relatively common for data scientists.

If the operations team doesn't understand distributed systems, they may be unable to keep your distributed systems running. As a result, there may be difficulty in production. An ability gap for the operations team is worrisome because that will pull the data engineering team's resources into production issues instead of creating data products.

If the data engineering team is weak in distributed systems, you have a really serious problem. Not every software engineer will be able to make the switch from small data systems to distributed systems, because of the significant increase in complexity. The organization should stop the project until there are substantial fixes to the team.

Management will need to quickly figure out whether the data engineering or operations team has an ability gap or skills gaps. Has the team been given the necessary learning resources to understand the distributed systems they're expected to use? Not everyone on the team may be able to pick up a book and understand the distributed systems concepts. Different learning resources may be needed, such as in-person training on the technologies themselves, training videos on the technology, or attending a conference dedicated to distributed systems or the technology itself.

Time is another crucial factor. Management often assumes that the time to learn and understand distributed systems is the same as for small data. In my experience, it takes about six months for a person to start feeling comfortable with the distributed systems technologies they are working with. For some people, it can take as long as a year.

Figuring out whether a person has an ability gap on distributed systems will take time. Management will need to reassess staff over smaller periods to make sure the person is making improvements as time goes on. Otherwise, management will need to hire from outside or get a consultant to convey the distributed systems skills to the teams.

My team says that you're wrong and that the task is easy

There are definitely some outlier teams and staff. Usually, these are people who already have extensive backgrounds in data, multithreading, or massively parallel processing (MPP) systems. For them, moving to distributed systems is a lateral move instead of a massive increase in the complexity of systems creation. For people with these backgrounds, the learning and level of difficulty go way down. In these cases, you could have an outlier team: a team at a significantly more advanced starting point than other teams.

Another possibility is that the team hasn't fully understood or internalized the complexities associated with distributed systems. Given a cursory knowledge of distributed systems, they may seem easy. However, the team understands only a small piece of what they need. Once the full problem and extent of difficulty are comprehended, the team may realize that their task is more complicated and that a great deal of work goes into being successful with distributed systems.

Why can't I just hire some beginners and have them create this project?

This attitude among management springs from several misconceptions about the difficulty and work of data teams. These misconceptions need to be cleared up as soon as possible, or the team will face a high probability of failure.

One misconception is that management thinks that different beginners can complement each other's skills. Management believes that what one beginner doesn't know or understand, another beginner will know. The reality is that beginners will have the same knowledge, and there will be very low complementary understandings (<<junior_only.>>).

Another misconception concerns the increase in complexity for distributed systems. Management may have had success with small data projects by swarming beginners on a project, and they were able to get something into production. Management thinks the same sort of plan will be successful with distributed systems.

At the root of these misconceptions may come a more nuanced question: why does management want a bunch of beginners for the team? Is it because management wants beginners because they're cheaper than an experienced person? Do they think that for the amount that the organization could pay one experienced person, they could hire five beginners? Managers may have the illusion that five people will have the same or much higher productivity than one person. In my experience, an expert with distributed systems will have at least 10 times the level of productivity as an intermediate person and as high as 50 times more than a beginner.

The reality is that a data team will skew to more senior people. It can have people at all levels of skill, but a team of just beginners will have severe difficulties with productivity. I've seen too many beginners on a team sap the productivity of the few senior people with too many questions and constant misunderstandings that need to be rectified.

We built out a distributed systems cluster, and no one is using it

Some organizations think that if you build a cluster, people will just start using it. This doesn't happen, and after a while, the organization begins to wonder how they can begin to benefit from the data products.

First, people need technical access to data. Even with a distributed systems infrastructure, data may still be hidden in silos. To create value, the data needs to be transferred to various places around the organization.

There may also be a skills gap for the rest of the organization. While the infrastructure was one part of the problem, it wasn't the entire problem. The teams will need to spend time and resources to really utilize the new distributed systems.

Management often makes the mistake of expecting the teams to learn distributed systems over a single weekend and come in with all of the necessary skills. There is far more involved in skill acquisition. Don't attempt shortcuts to what is a long and challenging journey to becoming a data-driven organization.

Out-of-Whack Ratios

Some organizations are entirely lacking or have very few of one of the required data teams. In these situations, staff are unable to do their jobs because they waste time doing things for which they aren't qualified: data scientists trying to do data engineering or operations, for instance. This happens when management doesn't follow the best practices with data teams or doesn't realize the different skills needed for data teams.

My data scientists keep complaining that they have to do everything themselves

Most data scientists expect to spend the majority of their time on the trappings of data science: analyzing the data and developing models. When they can't, management will start to hear complaints from them. They will complain that they are spending their time on data engineering or operations tasks instead of doing what data scientists thought they were hired to do. This usually stems from one or more of the following:

- The ratio of data engineers to data scientists is wrong. The organization could be completely missing the data engineering team. Or the proportion of data engineers to data scientists may not be keeping up with the ambitions of the organization, and there may not be enough data engineers to keep up with demands.

- Some organizations will think that they have data engineers or have the title of data engineer with people in the role. However, the data engineers may not adhere to the definition of a data engineer, as put forth in this book. These data engineers may not have the skills to create the necessary systems and code. As a result, these duties are falling on the shoulders of the data scientists.

- For some organizations, the two teams may not be working together well due to some kind of political animosity or organizational strife. These issues could be keeping the two teams from working well together, and each side is feeling like it has to do all of the work.

- The data scientists may not understand the role of a data engineer or the complexity of what needs to be done when creating the data products. The actual amount of work the data scientists are doing will be a small portion in comparison to the real work needed to develop the data products. The data scientist's more modest contribution to the overall product seems much more significant in their eyes than what they're actually contributing as part of the overall data product.

- There could be significant problems with the data products being created by the data engineering teams, rendering the products unusable by the data scientists. In response, the data scientists will start to create their own data products to insulate themselves from the poor work of the data engineers. These data quality issues will make the data science team lose all faith in the data engineering team's ability to create usable data products, and the data scientists will continually feel forced to create their own.

When data scientists feel they are spending too much time on data engineering tasks, they will eventually quit and seek another organization that has a better data engineering culture.

My data scientists keep dropping or stopping projects

At some organizations, the data scientists silently fail or stop working on a data project. The data scientists could have various reasons for stopping projects, some correct, and some not. Not every data science project will be possible, and those projects can be ended for good reasons without results. But when a data science team stops a project due to a technical limitation, the project is finished incorrectly. The technical limitation may have been insurmountable for the data science team but achievable with a good data engineering team.

When data scientists are dropping projects due to technical limitations, the problems may be

- The organization is missing a data engineering team entirely or doesn't have the right ratio of data engineers to data scientists. If the data scientists don't have a data engineering team to get help from, they will hit the technical limitation and just stop.

- Some organizations will have data engineers as a title, but they don't meet the definition as set out in this book. As the data scientists seek advanced technical assistance from the data engineering team, the data engineers will be unable to provide the necessary help or support due to their own technical limitations. This leaves the data scientists to fend for themselves on data engineering tasks.

The data engineers' analysis isn't very good

The data engineering team should have people with some analytical skills in order to create a report or dashboard. For a data engineering team, the required analytical skill is rudimentary. It may include simple math, such as counts or sums. It may be helpful to provide the data engineers with some basic knowledge of data science to increase the data engineers' analytical abilities.

Project Failures and Silver Bullets

Some organizations move from one technology to another or one buzzword to another. They're in search of the next big thing, and they always move on without achieving their goals. As the organization fails, they consistently blame the technology for their failures and mistakes. For these organizations, it's a challenge to look inward and fix the management reasons for the various shortcomings. But until the organization fixes its management issues, the teams will never be able to create a successful project.

Some organizations seek silver bullets. They believe that a technology bet will save the company or create such a massive ROI that will solve the organization's financial problems. Although big data projects can create enormous value, they probably won't save a failing organization.

We've tried several projects, and none of them went anywhere

When several data projects have gone nowhere or didn't show any value, the organization starts to look for answers. The easy fallback solution is to start blaming technology for the failures. The much more in-depth and fundamental questions are more challenging to find and fix. The management team needs to start looking deeper after several project failures to figure out the real reason or reasons why things went wrong.

Management needs to ask whether they really set the team up for success or whether there was just some ticking off of boxes. Did the team have all of the required resources, or were there only a subset? Did the management team just designate unqualified people as data science, data engineering, and operations teams?

Sometimes management doesn't understand the increases in complexity created by big data projects. They equate creating data products with their much simpler projects such as web development. As a result, they think data projects can be finished in the same time frames as other, comparatively more straightforward projects. In these scenarios, the project is canceled after a month when four to five months are necessary. Management needs to review that the team was given enough time to finish the project and ramp up on the technologies if necessary.

We've given up on big data because we can't get any ROI

Big data comes with a lot of baggage and overhead. An organization may not realize or plan for the extra complexity of big data. Some organizations will come in expecting the more significant ROI promised by data projects along with the much lower investment required for small data projects.

In these situations, the issue is probably due to organizational problems and not technical problems. More than likely, it's one or more of the following issues:

- The organization is missing one or all of the data teams. The organization hasn't internalized the need or value created by each team. They think they can get by with just one or two teams instead of all of them.

- The teams are missing people with the actual skill for the role. For example, the data engineering team could be missing the distributed systems and programming skills to create data pipelines. Without the ability to create a data pipeline, the organization will never be able to generate ROI because the data doesn't exist.

- Another issue could be that the organization is cheaping out or paying bottom dollar for people, technology, or help. Some organizations try to pay 20 percent under market for their people. They've been successful enough paying cheaply with small data projects, but that doesn't work anymore, given the increase in complexity due to big data. Other organizations try to get help, but seek out the cheapest source of guidance—which in turn are paying their own people cheaply—and the cycle repeats itself.

- Sometimes, the people are not set up for success. If a team is brand new to distributed systems or data, they will need the various resources set out in this book to succeed.

- For more complicated use cases, the teams could be lacking the veteran skills and leadership they need. Without these skills, the rest of the team can't plan or conceptualize what to do in order to finish the data product.

- The business side could be dismissing or not putting in the effort required to be successful. The business may have created an amorphous data strategy that lacked any real detail or body. This was given to the data teams to implement, and the business washed their hands of any more input. Instead, the business needs to continue to work with the data teams to verify that the data products really do meet the business needs.

- Organizations that lack a clear business goal or outcome won't ever hit a goal. Without an end goal, the team can't create any ROI. The business and technical teams need to identify a clear business goal and figure out how data products can achieve that goal.

We've tried cloud and failed; now we're failing with big data

Some organizations use technology as a crutch to deal with more fundamental issues. They see each new technology as a silver bullet that will change the fortunes of the organization. These organizations hop from one technology to another, expecting or buying into the technology vendor's hype that it will dramatically improve their business.

The organization never addresses the root issue of why the organization keeps on failing. As each new technology is added, the organization goes through the motions for another failure. The previous management or organizational issues weren't fixed, and the same issues will repeat again with a new technology. Although the organization points the finger at the technology as the reason for failure, the root of the problems lies with management and organizational issues.

In these repeating technology failure scenarios, the issues are likely due to:

- The organization is missing one or all of the data teams. They've been told by the technology vendor or have assumed that the new technology makes everything easier. With this assumed ease, the management team doesn't think they need all three teams.

- The teams are missing people with the actual skill for the role. For example, the data engineering team could be missing the distributed systems and programming skills to create data pipelines. This prevents the team from ever truly succeeding with the technologies.

- Another issue could be that the organization is cheaping out or paying bottom dollar for people, technology, or help.

- Sometimes, the people are not set up for success. The organization expects that everyone learns the new technologies on their own time instead of being given the resources to succeed.

- For more complicated use cases, the teams could be lacking the veteran skills and leadership they need.

- The business side could be dismissing or not putting in the effort needed to be successful. The business may have chosen the latest technology as a silver bullet and then stepped away from all further responsibilities. Instead, the business needs to continue to work with the data teams to verify that the technology implementation really meets the business needs.

This is really hard; is there an easier way to get it done?

The short answer is no. The long answer is yes, but it won't be as easy as you hope. There isn't an easy button to push to ramp up data systems. Organizations seeking easy answers and technology implementations often fail or vastly underperform.

One option for an easier way to initial success is to hire costly consultants to come in and do everything. The organization would essentially outsource all of the data teams to an outside vendor, and the outside vendor will do everything for the organization. Finding an outside vendor for this task is complicated. It's tricky to find a consulting firm that's really up to the task. A corollary issue is that when management sees the bill for the competent outside vendor to come in and do everything, they get sticker shock. They start to look for cheaper alternatives that don't have all of the skills.

217

I've found that organizations that are so focused on ease of development have very low odds of success. In these cases, I think these organizations are better off not even attempting big data projects.

Management's focus on ease comes from their feeling that there is a technology out there that makes all of this easy. Let me share Jesse's law on complexity. Big data systems can be made less complicated only for a specific use case or vertical/industry. I don't believe it's possible for a general-purpose distributed system to be easy. I think certain parts can be made easier, but that is only relative—overall, that doesn't make usage easy.

At some organizations, the management team assumes that the lack of ease is due to the complexity of programming. If the team could just use technologies that don't require programming, they'd be on easy street. But programming is only one of the critical skills that big data calls for. There's a key reason I split out distributed systems as a separate skill: it is just as key as programming. The distributed systems are where the majority of complexity comes in, and there just aren't shortcuts to make general-purpose distributed systems easy. You might be able to put some best practices or cookie cutters in place, but you'll need a qualified data engineer for the foreseeable future.

Is there any simple way or shortcuts to these problems?

Yes, sometimes, there could be more straightforward solutions to problems. Finding simpler solutions to problems is a crucial place where your veteran skill comes into play. The value of the veteran skill is to keep the team—especially new data engineering teams—from doing something stupid. This could be solved by finding a more straightforward way to do the same thing. Also, note that simple is relative, and there won't be a time where management is marveling at how simple things are.

Some teams and organizations will try to do everything all at once, no matter how difficult the implementation is. I highly recommend breaking down projects into more manageable and more straightforward steps—namely, crawl, walk, run. By breaking down much more extensive and long-term tasks into shorter, more attainable tasks, a team will be able to create the velocity necessary to accomplish the challenging parts.

Teams that are brand new to distributed systems or big data can get easily overwhelmed. It can look incredibly convoluted from the outside when the team members start to look at what's required. Teams can't go from 0 (small data) to 100 (big data) with ease or without any growing pains. Your veterans should be there to help the team break down the complexity into more manageable parts and guide them through the overwhelming feelings.

We hired a consultancy to help us, but they aren't getting it done

I've found problems with consultancies to be self-inflicted—the organization created its own problems by not following best practices.

When the management team was evaluating consulting companies, did they go with the cheapest bid? To achieve really low bids, the consulting companies will either pay their people meager wages or contract out the work to an even lower-paying consulting company. In either case, the quality of people may be so low that they'll never get anything done.

Did the management team check for references and find other successful clients of the consulting company? While making the vendor selection, was there someone in the room who knew the right questions to ask? Did the salesperson just assuage any doubts, or was there actual substance to responses?

Another common tactic is that the consulting company will pull a bait and switch. During the sales cycle and initial meetings, the consulting company will put their most competent people forward. When the project starts and the implementation begins, the consulting company will send their less-qualified people. These less-skilled people won't have the ability to get the job done but won't admit any issues because that will get them fired. Depending on the contract, it may not behoove the consulting company to make any changes because any delays or issues are paid for by the organization instead of by the consultancy.

We followed everything our vendor told us to do, and we're still not successful

The sad truth is that many vendors are in it for themselves and not their clients. The vendors will tell their clients whatever they need to hear to make the sale. The salespeople are more worried about losing a deal than keeping a long-term client.

During a sales cycle, they will tell you how their products solve a problem rather than selling the right tool to solve the problem. The salespeople are also blinded by the need to map every technical issue to their solutions, no matter how poor a fit the technology is for the use case or how poor a job the vendor did at implementing the technology. Sadly, some clients depend on outside vendors to give them advice about technology when the vendors consider it the client's duty to make the right choice for themselves.

Other vendors are on a fishing expedition to find someone—anyone—to use their products. This is especially true for new products that a vendor creates. They're more interested in adoption numbers than real client success with the product. This leaves the client holding the bag when the product is ultimately unsuccessful.

We defined our data strategy, but nothing is being created

Some management teams think their jobs end with creating the data strategy. From their viewpoint, all of the work from then on is done by the rest of the organization. This is a fairly common misconception from management, especially at the C-level.

The importance of the business and management team continuously working with the data teams can't be underestimated. In this book, I dedicated an entire chapter to how the business should be working with the data teams.

Creating data teams isn't just a technical shift. An entire cultural shift needs to happen organization-wide. The data teams can bring about only limited culture change. The management team bears the brunt of making the required shifts.

Who was the data strategy handed off to? If it was a general IT function, the IT department may not have the right people and might have just dropped the request. This is a manifestation of the organization missing data teams. Without all of the data teams, the organization will be unable to execute on the entire data strategy.

Our data science models keep failing in production

Models are only as good as the data coming into them. If there is bad data, the data teams should put some effort into figuring out what is happening and how to prevent the issues. For example, there may need to be more error and data checking for the program.

The model itself could be having problems too. The data engineers may need to review the code for the data scientist's model to verify that it checks its data. The way that the model is called could be an issue too. If the data or data model is too unstructured or not well defined, the model could be failing due to coding problems.

The Holy Grail

Organizations in search of the technical holy grail are relatively common in big data. These holy grails represent the most complicated and sought-after designs and architectures that really advanced companies have created over the long term. This search is especially universal for organizations that are very new to big data and distributed systems, where they haven't fully understood or internalized the changes in complexity they'll face or how much the organization itself will need to change to maximize the value created. They just think they can copy what the other organization is doing and achieve the exact same results.

Sometimes, data teams are forced to promise holy grails in order to get funding. To compete with other budget allocations, the data teams have to promise unattainable or incredibly lofty goals. This leaves the teams biting off more than they can chew. These teams are never given the opportunity to walk, much less crawl, before being expected to run or sprint (see Chapter 7, section "Create a Crawl, Walk, Run Plan").

We copied someone's architecture, and we're not getting the same value

I've worked with a lot of organizations all over the world, some of them direct competitors. I had a deep dive into their architecture and code. Despite being in the same business or industry, their technology stacks and code were very different.

Why would their technical implementations be so different despite being in the same business? The key driver for data should be around business value and business goals. Each organization has somewhat different goals and therefore requires different implementations to achieve business value.

The technologies should be enabling specific use cases. The use cases between the companies will vary, and they will use different technologies.

Finally, the teams are different between the organizations. One organization will pay better and have more competent people that will be able to implement more complicated use cases. Another organization will pay worse and have people who struggle to implement complicated use cases.

While imitation may be the sincerest form of flattery, it isn't always the best route for creating architectures or choosing technologies. It's used as a shortcut by organizations that are trying to get ahead faster. The copied architecture may or may not be the right one for the organization, use case, or people.

In conference talks and white papers where organizations boast about their successful architectures shared, they rarely share some relevant but crucial information such as:

- The amount of outside help they've received while implementing the architecture. The presenter's organization may have enjoyed extensive support from an outside consulting company or a technology vendor's solution architects. This outside help may have given them a prodigious leg up while implementing the architecture they're showing.

- The starting experience of the team with distributed systems, programming, or the technologies in question. Did the team already have people with a deep background in distributed systems, permitting staff to make a relatively straightforward lateral move instead of struggling with a big step up in complexity? Did the team hire a person with extensive experience in the technology at a previous job, someone who brought a great deal of experience to the team that wouldn't have been there otherwise?

- The presenter usually leaves out how long they've been implementing the architecture. It's a rare organization that will honestly share its journey from the beginning to its current state. For some organizations with a holy grail architecture, the journey has been 10+ years. This fact may not even be mentioned in talk or white paper.

- Most technical conference talks and white papers focus on the technology or implementation. They leave out the nuances of the business case or don't even cover the business case at all. This forces the attendee or reader to decipher the business case that the technology or architecture enabled or the value it created.

- I've attended many conferences where an executive is giving a talk about their organization's journey. Often, I will have inside information or have firsthand knowledge about the real situation at the organization. What the executive says and what the reality is are two different stories. They might sincerely overstate their success or deliberately lie about how far along they are.

We have a really ambitious plan, and we're having trouble accomplishing it

When you envision a holy grail, the ambitious plans that result are challenging to achieve.

If an organization is just beginning with big data, the team likely lacks any velocity (see Chapter 7). This velocity will allow the team members to get some experience and confidence in distributed systems. Organizations that try a shortcut to velocity have significant difficulties, and the projects tend to go nowhere. The organization should focus more on attainable goals and on creating momentum for the teams.

Comparing one organization to another in terms of productivity can be difficult. One organization may be getting copious outside help, while another organization is slogging along and trying to go it alone. The organization getting lots of support may be appreciably more successful. Instead of comparing success rates, the management team should be looking at how to make their teams more successful with more or better resources.

Another issue could be too many beginners and not enough veterans. When companies that are just beginning with distributed systems create ambitious plans and assign their implementation to beginners, they are creating a recipe for disaster. There are two equally poor outcomes for these situations. One is that the team creates a solution that is held together with duct tape, hoping to meet the ambitious plans while making a solution unworthy for production use. The other possible outcome is that the team never makes any progress. Teams really need veterans to help them create production-worthy systems—especially when the architecture is particularly ambitious.

The Software or Data Pipeline Keeps Failing in Production

An organization may congratulate itself that their job is done when they deploy their code to production. One of the many manifestations of complexity with distributed systems turns up in the difficulty of operations. Instead of the problems being relegated to one system, the issues are scattered throughout a cluster of hundreds of processes.

Many software products and architectures work in theory. The real crucible is when the code or architecture is put into production. That's when the really difficult operational questions are answered. Does the system scale to handle the load? Sometimes, the answer is no, and the organization has to firefight a data pipeline that is failing in production.

We keep on having production outages

Systems that worked in development often fail when put into production. This can have several different root causes.

Is the organization missing the operations team entirely? Who in the organization or team is responsible for the software, distributed system frameworks, and hardware running in production? If no one is ultimately responsible, the organization is missing an operations team and needs to create one.

Some outages are relegated to specific areas. Is the disruption related to or specifically coming from a new distributed systems framework that the data teams recently deployed? If so, what level of resources were given to the teams to be successful? An operational outage could have a code or architectural root cause, instead of a traditional operational issue. If no one was given the right resources such as training, the data engineers, data scientists, or operations team may not have enough knowledge to have used the technology correctly.

An operations team could be missing the troubleshooting skill. This skill is crucial to finding the really gnarly issues with distributed systems that take time and effort to track down. For some recurring production issues or outages, it will come down to the operations team's troubleshooting to figure out a deep or hard-to-find problem.

If an organization is in the cloud, there could be issues related to the cloud provider. On the cloud, all resources are shared. Sometimes a single node in a distributed systems cluster causes intermittent problems. One such issue is called the noisy neighbor problem. The operations team will want to keep an eye on not just overall cluster performance but individual node performance too.

The data keeps on bringing our system down, and we can't stop it

One of the significant difficulties of data pipelines can be the data itself. Problems with data can be incredibly challenging to figure out and fix. The issues can affect or come from all the data teams simultaneously. This can put a strain on communication when the groups have to triage an issue.

On the operations side, does the operations team really understand the data coming through the system? Do they know the data format and what a correctly formed data format looks like? Does the operations team have tools to inspect the data for the correctness, or is there a really convoluted way to check on data?

On the data engineer side, procedures for dealing with bad data need to be written into the code. If the code isn't dealing with bad data, the team may not be coding defensively enough. The data engineer's code might be expecting the data to flow cleanly, and there could be times when the data doesn't conform to these expectations. So the data engineer's code should be checking for anomalies. There should be adequate unit testing and integration testing that not only tests for a happy path (good data) but to make sure the code correctly flags bad data.

Data pipelines can have various stages of cleanliness. There could be a raw data pipeline where the data is saved or moved around precisely as it was received. This data will have all kinds of errors and incorrectness. There should also be a clean data pipeline for data previously processed by the data engineers to remove any bad elements. This data should be entirely made up of good data, be as clean as possible, or have any anomalous or bad data flagged. Unless there is a substantial need, the majority of the organization and use cases should be using the clean data pipeline. By mistake, a program or person could be using the wrong data pipeline.

The data scientist's code could be the source of data issues. Their code could be creating bad data, or their code could be failing due to the bad data. It's important to review data scientist's code because of their lower sophistication with software engineering and programming in general. The code should be checked to verify that it is defensively checking for correctness instead of assuming all data is correct.

Some architectures will make use of distributed databases or will use unstructured data. Some teams will go too far in their use of unstructured or string-based formats because they perceive that these formats will give the highest level of flexibility or extensibility. With high levels of flexibility comes great responsibility. Teams may want to move to a binary data format to prevent the many associated issues with unstructured or string-based data formats.

It takes way too long to find and fix issues in code and production

Finding issues in production can be the symptom of a myriad of deficiencies at the operational and development levels. Adding to the difficulty of finding the problem can be an intense pressure from the business to get back up and running. These trying times can really bring a data team's productivity to its knees.

If operations can't find the source issue, they may lack enough monitoring or logging to find and identify problems. The operations team should have extensive monitoring systems to quickly find a flawed process or node in a distributed system. The operations team should also have adequate logging, along with systems to search through logs. Without these systems, the operations team will have to search through too many systems to find the culprit. Management should make sure that the operations teams have all of the resources they need to find operational issues quickly.

Once the issue is found, a common problem here can be that the data engineer team is lacking unit tests, integration tests, or performance tests. These are an integral part of good software engineering, and hence of the data engineer's toolkit. A lack of unit testing forces manual checks for problem regression, forcing the data engineering team to take a while to fix issues.

PART IV

Case Studies and Interviews

I want conflict I want dissent
I want the scene to represent

—"The Separation of Church and Skate" by NOFX

In between books with ivory tower scenarios and theoretical approaches lies the real world. Although this book represents my experiences in creating and working with data teams, I didn't want the book to have only my thoughts. I want to share the ideas of the people who are out there dealing with the messy real world on a daily basis. These are their stories.

Some of the opinions and experiences will go directly against or will appear to go against the recommendations I've made in this book. I decided to keep these passages in the chapters to show there are several different starting points and viewpoints on data teams. There are a plethora of reasons why there is a difference. I made the decision not to editorialize on what the interviewees said.

All of these interviews represent the views of the individuals and not necessarily those of the companies where they worked.

CHAPTER 14

Interview with Eric Colson and Brad Klingenberg

About This Interview

People	Eric Colson and Brad Klingenberg
Time period	2011–2019
Project management frameworks	Custom
Companies covered	Stitch Fix

Background

Stitch Fix is the world's leading online personal styling service—combining data science and human judgment to deliver apparel, shoes, and accessories personalized to clients' unique tastes, lifestyles, and budgets. Stitch Fix is available for women, men, and kids in the US and now for women and men in the UK; Stitch Fix's goal is to help its clients look, feel, and be their best selves.

Eric Colson started Stitch Fix's data team in 2012. His official title was Chief Algorithms Officer, and he has now transitioned into an Emeritus role. Before moving to Stitch Fix, he was the Vice President of Data Science and Engineering at Netflix and was an Analytics Manager at Yahoo. Eric has a bachelor's in Economics, a master's degree in Information Systems, and another master's degree in Management Science and Engineering.

© Jesse Anderson 2020
J. Anderson, *Data Teams*, https://doi.org/10.1007/978-1-4842-6228-3_14

Brad Klingenberg is the Chief Algorithms Officer at Stitch Fix and is Eric's successor. Brad was an early member of Stitch Fix's data team. He's worked as a data scientist at other companies such as Google, Netflix, and other financial companies. He has a Ph.D. in Statistics from Stanford.

Starting Point

When Eric came to Stitch Fix to create the Algorithms department, he started from scratch. At that time, the company's technical stack looked like a typical Ruby web development stack, with a Ruby web application, a database, and web servers. All of the data was coming from the web team or from internal systems (inventory management, etc.).

In a typical fashion for web developers, many fields would be overwritten whenever there was a change. For example, whenever there was a price change on a piece of clothing, the previous and historical prices weren't saved. This story highlights one of the big differences between web developers and members of data teams. If it were up to the data scientists, data would never be deleted or updated. It also shows some of the early growth and changes that organizations have to go through. In order for the data scientists to start creating pricing models, the web engineering team needed to create a historical dataset of all prices.

One big advantage the web developers offered the data scientists was a high-quality data model. Stitch Fix made the integrity and veracity of data an early priority. The data was exposed through a relational database with an intuitive and consistent schema. There were a few mutability concerns in the relational database usage that had to be changed, such as not keeping historical pricing as this was simply overwritten. Aside from the mutability concerns, this high-quality data meant that the data scientists could focus on value creation instead of trying to decide if they could trust the data. When there are anomalies in data, the data scientist can focus on figuring out the business reason for an anomaly and how to exploit it instead of trying to figure out the technical reason why a piece of data is flawed. The data was all stored well in the database, and the data scientists could start using without huge modifications.

Growth and Hiring

At first, Eric was doing everything himself. He quickly found the need to create a team. His first three hires were data scientists. The first hire initially focused on providing basic insights into the business. The second was hired to work on the merchandising areas. Brad, the third data scientist, was hired to work on the styling algorithm and improve on some of Eric's early models.

Since the company was new, data volumes were initially tractable. The Algorithms department didn't have to do much data engineering because they were able to piggyback off the systems the web engineers had created. Thus, the data scientists replicated the PostgreSQL database the web engineers used and created their own instance. From there, all analytic processing could be done in memory and with a single computer using Python or R.

As the team grew to 20–24 people, Eric and the team had to grapple with how to create an organizational structure. They didn't just want to re-create the default management or organizational leadership strategy that was the norm at their previous companies. Instead, the team held off-sites to start with the fundamentals of building their department. They debated topics like "what is the role of hierarchy?" and "what are we optimizing for when we design the org?". They decided to build an organizational structure optimized for innovation.[1] They were quite explicit in this and recognized that this meant they would not be as good in other areas like system stability or executing on prefabricated requirements. But they knew that sacrificing in these areas was well worth the benefit of more innovation.

In line with that goal, they instituted a culture of hands-on managers who liked to get their hands dirty with data science work instead of mostly administrative work. Each manager had just a few direct reports—initially capped at about five people—to allow the managers to keep a sense of accomplishment and participate intellectually in real data science capabilities. Brad was one of the first hands-on managers and continued to focus on the styling part of the business.

Just creating a management structure doesn't guarantee productivity or accomplish organizational goals. They coined the phrase, "delegate accountability with confidence," to justify the introduction of hierarchy. This means that the people management roles exist to own an entire algorithmic capability (or set of capabilities), including all the

---

[1]See more about how Stitch Fix cultivates and grows data science at `https://cultivating-algos.stitchfix.com/`.

technical functions (modeling, ETL, implementation, measurement, etc.) as well as the people functions (hiring, managing, providing feedback, etc.). This vertical focus on capabilities, rather than a horizontal focus on technical function, allowed the team to scale up to over 100 data scientists while preserving the autonomy needed to move quickly. Further, the organizational structure allows them to deliver on business results while fostering bottom-up innovation.

In practice, this approach meant that data scientists had to be diverse in skills to manage an entire capability. They had to become generalists and are expected to do their own modeling, ETL, and deployments. This was necessary to avoid hand-offs and scheduling challenges that are associated with a more specialized division of labor. The capabilities became the basis for organization. Related capabilities are grouped together to create teams, and groups of teams create the department. Grouping the people this way created autonomy for each team as it rolled up into the rest of the organization.

Eventually, the department grew big enough to need another layer of management. At that point, Eric hired a couple of director-level people to oversee the various teams. This allowed the team to continue to scale. Once Brad took over the team, he scaled things even further, and today the Algorithms team has over 125 data scientists and platform engineers.

The Primary Division into Data Science and Platform Teams

Eventually, doing all the processing in-memory with Postgres started hitting the limitations of the database and server hardware. The team had to deal with petabyte-scale data and needed to develop a custom platform for algorithmic processing. Further, they needed to provide clean interfaces (APIs) for delivering algorithmic results to the various production systems. These activities require the skills of computer scientists more than they do data scientists. As such, the Algorithms department at Stitch Fix is composed of two major groups: *Data Science* and *Algorithms Platform*, both reporting to the Chief Algorithms Officer. The Data Science team mostly conforms to the definitions as set forth in this book, although with a higher programming ability and more operations. The Platform team builds infrastructure and tooling to enable the data scientists to be autonomous.

Data Science Team

The Data Science teams are further divided into centers that align with the major functions of the company.

Merchandising Algorithms

This center develops algorithms to manage inventory and makes decisions about how to carry and create the best products for clients.

Styling Algorithms

This center focuses on algorithms that power the styling recommendation engine. More generally, the center handles a suite of algorithms that match clients with products, including outfit recommendations, stylist selection algorithms, and inventory targeting.

Client Algorithms

This center focuses on personalizing the experience of clients on the service, independent of the clothes themselves. For example, there are algorithms that personalize when and how to engage with each client.

Operations Algorithms

This center focuses on algorithms that pace the flow of resources throughout the company, ensuring that the supply of labor (warehouse workers, customer services agents, stylists, etc.) matches demand.

Customer Service Algorithms

This center focuses on algorithms that manage in-bound customer engagement (routing tickets, recommending responses, etc.).

The data scientists are responsible for writing their own ETL, training models, and deploying them to production. Stitch Fix wanted their data scientists to be as close to the business problem as possible—to work with the business to verify that a model or new approach actually solves or improves metrics. In some cases, the data scientists **are** the business, with responsibility for direct impact on revenue and other metrics.

Platform Team

The Platform team sets up all of the infrastructure and tooling necessary to develop and run algorithms. Effectively, they are relieving the data scientists from having to understand many of the advanced computer science concepts needed to run their algorithms—containerization, distributed processing, automatic failover, and so on. The data scientists can then write code at a higher level of abstraction, happily unaware of the distribution and parallelization happening behind the scenes. This keeps them focused on applying science to business problems while also avoiding hand-offs to others.

By hiring the right people for the data science and platform teams and letting each focus on what they're good at, managers allow the data scientists to be far more productive. Using the platform, a data scientist can own and create a solution for the entire business problem. Describing the platform, Eric said, "It really enables the data scientists to own every aspect of the problem that they're working on—from thinking about how to frame the model to getting the data, to designing algorithmic, training models, causally measuring impact, and working with partners across the business without having to surrender or hand-off work to another function along the way."

Operational responsibilities are shared between the platform and data science teams. Stitch Fix has instituted extensive monitoring of production systems, which helps them quickly identify the source of issues. Depending on what the apparent source of the issue is, the team responsible for the system is alerted to the need for operational help. Each algorithm has a different SLA depending on its maturity and risk factors. The team did not want a one-size-fits-all SLA since, in many areas, they valued innovation over stability.

The Stitch Fix architecture runs entirely in the cloud. The primary benefit they get from cloud computing is agility. Data scientists and platform engineers are able to spin up new clusters on a whim to try out a new idea. If the idea doesn't produce the desired impact, those resources can be spun down just as quickly. This mitigates the need to put a lot of processes around which ideas to try.

Bottom-Up Approach

I found Stitch Fix's approach to creating algorithms fascinating. At most organizations, there is a heavy top-down approach to things. A CxO makes some kind of mandate, and everyone executes on that mandate, or a business unit funds work on specific functionality. Stitch Fix, in contrast, has encouraged an emergent or bottom-up approach to data science.

"It's inherent to the fact that these types of algorithms can't be designed upfront; they have to be learned as you go," Eric shared. The company understood early on that the data science workflow is different from other disciplines. "Instead of specifying a model up-front, you have to let the data and ML *reveal* the model." The business can't just build a spec and expect the data scientists to create it.

Because of this open approach to learning, the data scientists themselves are the ones who find the need or possibility of improvement or optimization through a model. "The people that are closest to the data, that are working with the data day in and day out, find things that others won't. Many of our algorithmic capabilities were not asked for," Eric said. "The ideas didn't come top-down from a business person or even from myself or Brad. The ideas came from the data scientists themselves."

Eric claims that it's a curiosity that compels the data scientists to explore an idea. "They see relationships in the data that suggests a way of doing something more efficiently, or to build a whole new capability altogether," Eric said. "They can't help themselves. Once the observation is made, they are compelled to chase it down." With access to the data and virtually unlimited compute resources provided by the platform, the data scientist can explore those ideas nearly for free. "I call it *low-cost exploration*. They don't have to ask permission to explore those ideas."

But this curiosity isn't just supposition; it's backed by the data and statistics. "They have evidence - AUC (Area Under Curve) and other statistical measures - that let them know when they're on the right path. Not all ideas pan out to be good ones. But the evidence afforded by data and statistics can inform you as to whether you are going down a dead-end vs. on to something potentially game-changing." From there, the data scientist can bring their findings forward to the rest of the team to show the potential impact of their exploration.

Brad said, "Most people on the team are folks who are excited about solving business problems and finding ways to apply data and algorithms to the business. And in many ways, I think the biggest wins we have are the breakthroughs that come from people rethinking how we operate different parts of the business to use data and algorithms. This is as much about solving business problems as about solving math problems."

When we were talking about this emergence, I assumed Stitch Fix was doing something similar to Google's 20 percent project time policy, where each person can dedicate one day out of five to projects that aren't directly related to their day job. Stitch Fix decided not to do this as they were worried that too much structure and pressure to create something might actually be a detriment to innovation. They worried that each person would feel the need to only work on "earth-shattering" ideas when many simple ideas could be just as effective.

Instead, Eric and Brad decided to continue to let innovation be driven out of curiosity. "Data scientists are almost plagued by curiosity. That itch has to be scratched," Eric said. "I recently chatted with a few people on the team, asking 'what made you come up with that great capability that no one was asking for?' Their answer is almost always 'Because I had to.' 'I had to figure out why it wouldn't factorize.' Or, 'I had to explain that anomaly,' or 'I had to see if this hypothesis was true.'"

Although curiosity was the main basis for their research, the data scientists still required a good foundation of understanding about the business. Eric said, "Our data scientists are very well equipped with business context so they can judge for themselves how valuable something will be. They have access to the data to vet their ideas, and they have the business context to know how much impact a new capability can have."

The data scientists weren't the only ones with the ability to let a new idea emerge. The platform engineers are also driven by observation. Rather than take requirements from the data scientists, they observe how they work and build frameworks that will make data scientists far more effective—even if they are not asking for it.

Project Management

It's difficult to find a project management framework that's well suited to the research and development component of data science work. There are just too many twists and turns that the team can't anticipate and won't be able to plan for. Because of this, Stitch Fix doesn't use project management frameworks like Scrum or Kanban. They do something that I would describe as closer to open source project practices as described in Eric S. Raymond's *The Cathedral and the Bazaar*.

By having small teams and hands-on managers, the team doesn't need a great deal of project management structure. Instead, they can move quickly and informally. Brad said, "The approach we've taken is much more like being a gardener, you just want to create circumstances where people can do good work and, occasionally you need to trim a branch back or make room for a new sapling, but generally you're just trying to get the conditions right to then get out of the way."

Eric added, "The process has been miraculous, and very different in my view from how most data science companies are run today, which are much more top-down." In other top-down companies, someone outside of the data science team dictates or ideates, and then people from the Data Science team are assigned to execute on the idea. "The best ideas are going to come from the data scientists. They are the ones closest to the data. They see patterns and anomalies that the rest of us do not."

The Competitive Edge of Data

Stitch Fix finds itself in an insanely competitive segment of retail. They rely on data to maintain a competitive advantage. "The Stitch Fix business model provides a very different customer experience. But it's the data that is afforded from this model that provides us with our differentiation. It allows us to personalize better than other retailers," Eric said. Stitch Fix focuses on creating positive feedback loops and insights that allow them to serve their clients better.

With a brick-and-mortar retailer, many potential feedback loops aren't or can't be created. If someone leaves some clothes in a dressing room, the retailer won't track that the clothes were tried on and rejected. Ecommerce companies have more data about what's happening in their business but are only able to learn from items the customer has picked out. Stitch Fix focuses on closing these feedback loops. For example, Stitch Fix is able to get clients to try clothes that they wouldn't have picked out themselves. This allows them to capture preference data on fit, style, size, and so on—even if the client ends up returning some of the items. The data on why someone didn't like an item is often as valuable as data on why someone did like an item. This data helps Stitch Fix to recommend future items to clients as well as informs them as to which inventory to procure and even enables them to design new clothes.

Stitch Fix has a mutually beneficial relationship with its clients through data. Clients provide Stitch Fix with valuable data about their size and style preferences; Stitch Fix provides the clients with extremely relevant clothing—even clothes that wouldn't have thought of before. This symbiotic relationship through the use of data allows for both sides to benefit and creates a great reason for clients to share their true opinions. This would be hard to replicate by retailers with a more traditional model.

By having a true competitive edge to data and actually leveraging it, the algorithms team has a direct impact on the business. "I think something that is novel about our Algorithms Department is the amount of direct impact it has on the business," Eric

237

said. "Brad's department is accountable for real impact to various metrics, whether it's increasing revenue or optimizing to reduce costs. In addition to their own impact, they also partner with nearly every department in the company - marketing, merchandising, operations, styling, etc. - to provide algorithmic capabilities embedded in those functions. That is pretty unique to Stitch Fix: that real sense of accountability."

Stitch Fix is well known for its use of "human in the loop" algorithms. In some cases, they need the power of machine learning, combined with the nuanced abilities of human judgment. An example of this is the styling algorithms used to pick out clothes for a client. However, in other cases, there is no value added by human judgment, and all that is needed is machine computation. They call these "machine-to-machine algorithms." For example, how to divide inventory to various warehouses or how to match clients to stylists. These algorithms can start out as machine-to-human, but once there is enough confidence in the algorithm, the human can be completely removed from the loop.

Getting the level of trust that allows an organization to buy into machine-to-machine algorithms can really be tough. Brad said, "It's art and science - humans and machines working together. Sometimes there is tension, but it's a healthy tension. The goal is not just to automate something a human could do, but to equip the algorithms to do things that a human couldn't do by themselves. There are very interesting cases, sometimes where machines and humans disagree. That presents the challenge of earning trust and credibility. But over time, it's great when we can demonstrate that machines can create really good outcomes. Especially when empowering people in a company like Stitch Fix, you can make that case empirically through experimentation and using data to show that this actually makes clients happier."

Eric said, "Our business model really lends itself naturally to experimentation. We run randomized controlled trials where you get to know the causal relationships between things. This allows us to learn the outcomes of our decisions. And, it turns out that this can be quite humbling! We've learned that our intuitions fail us quite often. So many of our great ideas turn out to be wrong; they don't move the needle at all or can even move the needle the wrong way!". True to data science form, the team would sometimes write out their predictions on the outcomes of various experiments. After the experiment concludes, they would take the actual outcome and compare it to their predictions. "Often, we learn that not a single one of us was even directionally correct. It is sobering how far off we can be. But we need to learn this. It's likely that it's always been this way throughout business history. It's just that prior to having the ability to run properly controlled experiments, we never knew the outcomes of our decisions. Ignorance was bliss!"

Advice to Other Companies

I've talked to management all over the world who claim that companies like Stitch Fix can do what they're doing because they're in San Francisco. The managers assume that companies in the rest of the world can't leverage their data, hire well, or execute at the same level. I asked Eric and Brad this very question.

"I do think jobs and data science generally are not evenly distributed geographically. But there are growing hubs beyond just the Bay area or New York. The rise of more remote work is probably a good trend for people around the country as well," Brad said.

"Smart people are attracted to other smart people. This allows talent to amass, which, in turn, attracts more companies -- particularly the ones that really value top talent and offer the more innovative roles. The virtuous cycle concentrates the talent into relatively few places. So yes, it is a challenge if your company is not in one of those places," Eric said.

To build on that challenge, I asked them to give advice to the large enterprises that move at a snail's pace.

Eric said, "Well, I don't want to be too defeatist, but it's an uphill climb for an old, stodgy company. Companies tend to be resistant to change. Processes and values are established in the formative years and become inherent to the organization, acting sort of like a DNA for the company. If that DNA didn't include empiricism, experimentation, innovation, etc., it will be difficult to successfully introduce a data science team. It's like injecting a foreign entity into the body; defense mechanisms will reject it. It's not impossible, but it will be tough to make such fundamental changes to an existing company."

Brad advised, "Try to find a place in a company that offers both the data needed to solve problems and the appetite to use data. Ideally, you want a team that is not just using the data to get insights or to write a report but really wants to use data and algorithms to make decisions. And I think few organizations are really inclined in that direction. But within a large organization, there could be pockets of teams or capabilities where you have more opportunity."

Eric added, "Adding data science to an existing, mature company is a lot harder than having it from the beginning and allowing it to grow organically. As a new endeavor in an existing company, the data science team will likely have to declare it's initiatives upfront and even cite ROI estimates. By contrast, at Stitch Fix, data science was in the fabric of the company from the beginning. This allowed the data science team to explore opportunities and recognize the return on investment after-the-fact. That is, we would allow experimentation on many new ideas first and then invest in them with additional headcount after they are shown to be successful. This takes the pressure off of having to declare ROI estimates up-front on one single idea. Data science solutions have a lot of inherent uncertainty; it's best to explore widely and with a lot of trial and error."

Stitch Fix competes in a competitive market for people in the San Francisco area. They've focused on getting data scientists from a wide array of backgrounds such as chemistry, biology, and neuroscience and occasionally hire people right out of Ph.D. programs. To find the right people, Brad said, "working on a hard problem and really being trained to approach problems scientifically is a pretty general skill set. We generally find that people with a background in solving scientific problems and using technology and data do pretty well."

I asked how companies can benefit if they have significant C-level support for their data teams.

Stitch Fix already has this C-level support from early on. Eric was the Chief Algorithms Officer and was a member of the C-suite. "From the very beginning, we established data science as a first-class entity, with officer representation and a dedicated department," Eric said. Brad has since succeeded him into this role and is peers to the CTO and CFO.

Brad added, "Several benefits come from having everybody on the same team reporting up to somebody at the C-level. For the team, I think it provides us with a really remarkable community, along with better opportunities and career progression." By having Brad at the C-level, he's able to decide on how the data team's organization is structured. "And as far as the team reporting to somebody in the C suite, I think it's important to have a seat at the table. It means we are not just executing the ideas of another team, but really there to help shift the way the company thinks about data and uncertainty and using algorithms. As Eric mentioned, I think part of what makes Stitch Fix successful is the very empirical kind of data-first culture that we have. And I'd attribute much of that to, in particular, Eric being a C-level executive for many years after he first joined the company."

Lessons Learned in Hindsight

I asked what they would change or have done differently with the benefit of hindsight.

Eric said, "We caused some pain as data scientists rapidly developed models. The models got into production, but maybe they weren't coded in the most robust way, so we ended up having to replace them after they were up and running. So certainly, if I knew then what I know now, we would code all our algorithms up the right way at the start so we wouldn't have to rewrite them. But that is easy to say in hindsight!"

Brad claimed that he wouldn't change much. "I think the general principles that have served the team so well could work in quite a few environments: favoring bottom-up innovation, hiring generalists who are good at framing problems, and really trying to let people own things."

I asked Eric and Brad where they thought they did a really good job.

Eric said, "We made such a bold bet on algorithms. It proved to be so important to our business model. Much of our success is attributed to having a clean slate to establish things the right way from the beginning. You learn things in your career - often from your own mistakes - yet it can be very hard to correct them in-place. You often need a clean slate to start over. At Netflix, for example, I had a division of labor on the team between the analytics or business intelligence teams and the data science teams. Both reported to me, but their work was cleaved to leverage their separate skill sets. This resulted in fragmentation of context, tooling, and infrastructure. I wanted to fix this, but doing so meant undoing years of legacy code and processes, as well as having to turn over much of the team. This is particularly daunting to do while also delivering on company priorities. But, I had a clean slate at Stitch Fix. There was nothing to undo. I was able to establish the right organizing principles from the beginning. Of course, some things had to be learned as we went along. But, we were very careful to preserve optionality and avoid things that would be hard to undo."

Other good choices included "bringing in the platform team early to grow with us" and "setting ourselves up for some autonomy." Having the platform team in their organization allowed them to set their own direction and align on principles. Eric recalled, "we didn't want to have to 'hand-off' requirements from data scientists to an engineer for implementation. Hand-offs limit your ability to iterate quickly. Instead, our algorithms platform team built an infrastructure that enables data scientists to implement solutions themselves." This turned out to be a great move for them. Eric reminisced further, "I remember the very first version of our styling algorithm. It had been trained on historical data and looked very promising. When it came time to implementation, we had options: we could build the API to interface with the Engineering team, or they could do it. We ended up doing it ourselves, which was a great call. By owning the API, we could abstract the Engineering team from all of our experiments and iterations. If we hadn't built it ourselves, we would have had to negotiate to get on their roadmap for every change. This would have been very difficult as they had their own set of important priorities."

He finished, "The interfaces between the Engineering and Algorithms teams became pretty clear. In part, this was enabled by the fact that the two teams had completely different tech stacks: Engineering used Ruby; Algorithms used Python and R. This led to good decoupling, which has served us well."

CHAPTER 15

Interview with Dmitriy Ryaboy

About This Interview

People	Dmitriy Ryaboy
Time period	2011–2019
Project management frameworks	Scrum
Companies covered	Twitter, Cloudera, Zymergen

Background

Dmitriy Ryaboy started Twitter's data engineering platform, interned at Cloudera when it was a small startup, and is currently the Vice President of Software Engineering at Zymergen. He has a master's degree in Distributed Systems and Databases.

Hiring at Twitter

Dmitriy was the fourth person hired at Twitter to do analytics. At that point, there was an analytics manager and two engineers. The data engineering function grew out of the analytics department rather than the core engineering team.

© Jesse Anderson 2020
J. Anderson, *Data Teams*, https://doi.org/10.1007/978-1-4842-6228-3_15

At first, the team was doing it all: both the analysis and creating the infrastructure they needed. The team's manager had more of an analytic focus but could write the advanced SQL needed for analysis. The rest of the team were programmers who were being asked to do analytics. Some of the programmers already had backgrounds in analysis because they had been part of an analytics team before.

At those early stages, some analytics problems were difficult distributed systems problems, and others were more fundamental analytics problems. Dmitriy said, "The company fell into the trap of thinking, 'Oh, we'll get three deep learning experts, you know, and pay them oodles of money and they're gonna make magic.' And like, yeah but they have no data. What are they gonna do? The company had kind of mismatched expectations." Without the right data or systems, the data analysts would have been idle.

The fourth and fifth hires had strong analytics backgrounds. Twitter was starting to build its revenue platforms, an effort that was spun out of the analytics team. To increase the team size, Twitter acquired another startup called BackTape that focused on the confluence of engineering and analytics.

As the company requirements grew, people started to organically split into the two groups we normally associate with analytics and data engineering. This split started to happen after about a year. Dmitriy started to build up the data engineering team by looking for people who could help create the infrastructure. At this time, Dmitriy didn't have the actual title of manager, and nothing was really on paper yet. There was still a single manager for all of the analytics department. As the team grew, it became clear that there were too many people for a single manager to keep track of everyone. At that point, Dmitriy was officially promoted to manager of data engineering.

By 2012, Dmitriy's entire group was about 20 people strong. At this point, Twitter did a reorganization, moving the analytics team under the finance department and the data engineering team under the VP of Engineering. Dmitriy says, "It was important for my team to shift over into the core engineering team, even though this split engineering from analytics. On the one hand, it's really critical to have that good collaboration between people who are doing the analysis and people who are getting the data ready for them and figuring out different ways to present the data. But it's more important to have managers who are knowledgeable about engineering do things review and evaluate the data engineers. If data engineers are not part of the core engineering organization, they're not necessarily seen as much of an engineering discipline. The analysts don't necessarily have the right gauge for what's hard and what's easy. Whoever is closer to you in the organization tends to be closer in alignment with you mentally and culturally."

Dmitriy made three subteams out of the data engineering team. One team handled the traditional database technologies, another focused on the Hadoop computing engines, and the third was a platform engineering team that used Hadoop to build products and organization-wide reporting systems. The platform engineering team was originally in a different part of the organization, but Twitter decided to consolidate them with the data engineering team. By 2014, there were 50 people in Dmitriy's group. They also added another group to the data engineering team called Streaming Compute.

As technologies get adopted across the organization, it's important to think about who should own them. There is a difficult trade-off between stifling innovation and technology adoption on the one hand and trying to make people's jobs easier or use cases possible on the other. "You want to avoid pseudo platform teams who don't actually understand platforms, who's job success isn't dependent on delivering your platform, and whose day job is something else but they're moonlighting as a platform team. They should be platform engineers, and they should focus on that platform job."

Twitter had acquired a company called BackTape that created a distributed stream processing technology called Storm (later donated to the Apache Foundation as Apache Storm). Then Twitter acquired another company where Karthik Ramasamy worked. His team rewrote Storm as a new project called Heron (later donated to the Apache Foundation as Apache Heron). Heron maintained API compatibility with Storm, saving Twitter from having to rewrite a massive amount of code that invoked the Storm API.

In 2015, Dmitriy left the data engineering team to run the AB testing platform.

Challenges of Data and Analytics

On the surface, doing analytics on tweets seems easy. Once you add in the analytics requirement, things become more difficult for data engineering. Placing the data engineering team under the engineering team made it easier to have the necessary conversations about how to make software changes and the effect on analytics. Instead of finding out about a change at the last minute or not at all, the data engineering team could be part of the conversation about rearchitecting or making major system changes from the beginning.

This way, data engineering didn't feel like they were battling against the rest of the company to create data products. Instead, they were able to say, "I understand what you're doing, I understand what's happening. We need to change that; let's figure out together how that's gonna work on the data side."

There is a critical nuance to data products that doesn't exist as much in software engineering. As the system changes to a newer version, the data engineering team can't just forget the previous version of the data. They have to maintain archival compatibility for all versions of previous data. Dmitriy said, "Data engineers still have to deal with the mess that records from two years ago are in. Either you pay an enormous amount for a rewrite because you're dealing with petabytes, or you put some sort of shim in place. If you do the latter, you have to document the logic. And still, it doesn't matter how many years pass by; you're still going to have the weird data that looks different from newer data. So it's hard. And those conversations are much easier to have when you're a part of the same organization, and you're thinking about the whole engineering task as a complete feature set, rather than shunt things off and leave it up to the analysts to figure out."

Different types of issues arise when you have the analytics and data engineering functions in two different organizations, as Twitter did. Because the two teams have to continue to work closely together, they still have to maintain excellent relationships. "It definitely helps to maintain the human connection and remember that there's a person on the other side. There aren't just random tickets that show up, and you do them. Still, splitting up led to more formality. We got more serious, so the analyst can't just walk up to an engineer; they happen to know and ask for the SQL query to be changed or whatever. Analysts need to file a ticket. We have a workstream, and we need to prioritize, all of which introduces friction and slows down innovation. There was some frustration about that on the analyst side. They wanted to say, 'I know where the file is, all you have to do is this.' We'd have to say, 'Well, that's not really how that works anymore'."

Some of these changes in process aren't specific to data or to data teams; some just spring from company growth. "The old guard just kind of knows where everything is and why. And so they feel very empowered to simply change things because they can understand all the context. This causes the new guard to say, 'What the hell are you doing? We have 500 new things built on top of this capability, and you can just change it.' These issues of organizational distance and the resulting increase in complexity never go away. Hammering out these procedures is continual and always a work in progress."

Task Ownership Structure

The data engineering team had to become clearer about the separation of infrastructure teams and their responsibilities from the analytics teams and their responsibilities.

In Silicon Valley, most people will stay at a job for only a few years. This means that after two years, half or more of your team is completely different. Some of the people who leave have created systems or projects that have gone beyond trials and are in production. Dmitriy pointed out that changes in personnel force negotiations about, "Who is going to own it? At what point in debugging a problem is it okay to call in the data engineers?".

He went on, "Debugging those kinds of problems, I think, is where most of the friction came from. Ask the engineers what a problem is, and they'll shrug and say, 'Well, what are you feeding into the system that it doesn't recognize? The data is probably skewed, but I can't help you until you do a little more work.' And when neither side is really clear on where their responsibilities lie vs. the other side, that's where you get into messiness—especially when everybody's stressed and overworked."

To reduce the friction of overlapping responsibilities, the teams thought about "what ownership looks like and what support looks like. We identified specific needs and built some tools to enable them." Thus, the data engineering teams built reports to help troubleshoot jobs. They told the analysts where the logs for the jobs were so that they could read the errors for themselves. For every job, there was an owner field, and that's who would get any alerts. This helped clarify roles and responsibilities, so it was clear who should fix things when they got broken.

The Difficulty of Choosing Technologies and Creating Distributed Systems

Choosing to write your own distributed system is a really difficult choice. It requires a deep investment in technology and a great deal of expertise and specialization. Twitter has written several of its own distributed systems.

Twitter was using MySQL originally, and the database was bursting at the seams. The DBAs were getting unhappy about the amount of operational overhead they had to deal with. So they started to look at using Cassandra to replace some of the workloads that MySQL was doing. But when they realized that the main performance problem revolved around a difficult database operation called a compare and swap, they decided to create a new distributed transactional database system called Manhattan to handle the problem.

Dmitriy's team was one of the first production customers of Manhattan. They used it for counters (for instance, counting tweets) and monitoring. The team kept on iterating on improvements to Manhattan. "It's easy to get the 80% on that distributed system that lasts 20% as a ten year long out drag out sort of." This long tail issue of getting to the last 20 percent of issues is incredibly time-consuming and difficult. Instead of trying to be a general-purpose system, the Manhattan team "were very thoughtful about finding use cases where the particular subset of things that is easy to do now was sufficient and would provide value."

I asked Dmitriy what recommendations he would give about the choice between writing your own distributed systems and using an existing one. At that time, for large-scale real-time distributed systems, "There weren't any like Storm. But even Storm, once it hit Twitter scale, had some significant problems. It was good for us to rewrite it because nobody else was going to." Dmitriy thought about the alternatives to rewriting Storm and said, "We could have moved everybody over to Flink, but it's a pretty big migration and involves a number of issues."

The Storm and Heron rewrites weren't the only big changes. "We wound up sunsetting Pig, so we had to support only Scalding. That upset some people and probably lost us a bunch of cycles on the analytic side while people rewrote jobs, but we got through it. That allowed us then to explore the more modern SQL engines. we wound up pulling in Presto, and that's a big component within the company now."

"One big caveat: massive companies such as Twitter, Facebook, Google, Pinterest, and so on have different problems from smaller companies. The big ones solve problems for themselves and maybe open source their stuff, but it's built to solve their unique set of problems stuff. The tools may be solving problems at their enormous scale, but that doesn't mean those tools are right for everybody."

He went on, "I think you have to be realistic and clear-eyed about tradeoffs and benefits. Get somebody involved who didn't already spend seven years getting invested into some particular system, and have them help make that call. Then commit to it and follow through."

I like to point out to organizations the inevitability of having to switch distributed systems. Even a company with as much talent and resources as Twitter still didn't know everything or have a "one and done" for choosing a distributed system. They had to continue to iterate as technology, and business requirements changed.

Data Engineers, Data Scientists, and Operations Engineers

I asked Dmitriy his definition of a data scientist. He answered, "I think that a data scientist is somebody who is both curious and rigorous about diving into data and what the data can tell them. They come into their skills through a variety of ways. Some people are totally self-taught, while some have PhDs in machine learning or statistics or what have you. I think they must have to be interested in the business, and the value of that side of the question. Like, why are we asking this question? Sometimes it means doing some data engineering."

Dmitriy's definition of a data engineer is "a software engineer who happens to think about data at rest and how to best structure data so that it's most useful for answering questions and building data products."

Asked for his definition of an operations engineer, he said, "I think all engineers are operations engineers. It's just that some engineers are more focused on that than others. If you're responsible for code that runs in production, you're an operations engineer. If your stuff is live, a living and changing system, you're an operations engineer."

I asked what ratio he prefers between data scientists and data engineers. He said, "It depends on what kind of data scientists you have and what kind of data engineers you have." Some data scientists are heavily knowledgeable about engineering, whereas others just know R. "I think the ratio is changing because of what you can do with hosted platforms. Now you don't need as much effort to stand up that infrastructure, but you still need to know what you're doing."

Part of the ratio extends to the experience level for the data engineers. "You need people who are experienced and people who are really interested in the domain, instead of junior people who are interested and senior people who are experienced and can kind of navigate their way through and, and know what to watch out for in order to build that."

He offered, "Platform as a Service diminishes the need for operations staff. I think it has put even more emphasis on the architecture and programming side."

He went on to tell a brief story about hiring. "I interpreted some direction that my management gave me literally. They specified how many data engineers they wanted, but then said, 'Go hire data scientists—we'll tell you when to stop.' I did the classic thing of hiring a bunch of data scientists without increasing the number of data engineers. So guess what? Our data scientists were our data engineers. Naturally, they complained, 'Why am I doing this? I don't know how to do this. And I'm going to quit'."

"Of course," he explained, "the CEO isn't literally saying, 'Hire a data scientist.' He was trying to say, 'I want a functional data science organization. Build it.' We're in a better place now because we hired the engineers we needed. We have awesome data engineers."

Project Management Framework

Asked about project management, Dmitriy answered, "We use Scrum. I think big picture, so I believe that the framework you use doesn't really matter much. Go actually read the Agile Manifesto[1] and focus on that. Not the mechanics, not all the processes around it. The manifesto calls for 'people over process' and then teams go and rigidly define the Agile process. Don't worry about it. If the process isn't serving the purpose, throw it out. I think that applies to data science as well as engineering. I've heard the argument that you can't do data science that way, but you can."

Business and Data Team Interactions

It's important to establish KPIs, in order to allow a data team to figure out priorities and see whether they're achieving them. Dmitriy acknowledged the common technical KPIs around "predictability, stability, things like that," but the best KPIs for data teams concern how they're affecting the business. "Both data engineering and data science are there to create business value. So define their KPIs around business value."

The business always has more requests in the pipeline than data teams have the time to accomplish. Dmitriy's teams have a single funnel through which everyone submits their requests. "We meet with the head of the data science team as well as the project manager on the data science team, to figure out what the requests consist of and how that relates to other requests and priorities." From there, they plan out their road map for next year.

Keys to Success with Data Teams

Dmitriy says that managers need to "boost data validation up as high as possible. And that means collaborating with nondata teams, other engineering teams. That will set

---

[1] www.agilealliance.org/agile101/the-agile-manifesto/

yourself up for success instead of just running like a hamster in a wheel trying to like fix things. The other side of success, again, is focusing on business value. Don't solve data problems for the problem's sake."

Data teams management needs to be "a partner at the table where like the decisions are being made."

Management's job is to "build and smooth out relationships" with the rest of the organization. "If you are stuck not talking to the rest of the engineering team, and you're just focusing on one kind of problem, you might get good at that one kind of problem, but you have no idea what else is going on and why it's going on. That gap makes you a worse engineer."

He recommends that management have "a deep understanding of technology matters." Management also needs communication and leadership skills to be successful. "You need to be thoughtful, especially when you think about organization design and structure, these kinds of meta effects that take a long time to manifest that are not necessarily about the technology. The way you structure the team affects who is talking to whom. Your choice of structure determines where you're going to have friction or not going to have friction. And when the structure is not working, you have to be willing to change it. A lot of this is about getting the right people into situations where they'll have opportunities to hear and participate in the right conversations. Building trust can help the human connection."

CHAPTER 16

Interview with Bas Geerdink

About This Interview

People	Bas Geerdink
Time period	2013–2019
Project management frameworks	Scrum
Companies covered	ING, Rabobank

Background

Bas Geerdink has extensive experience in creating data-driven companies and using artificial intelligence in the financial industry. He used to manage data teams at ING Bank. Now he works at Rabobank. He has a master's degree in artificial intelligence.

> *ING is a global bank with a strong European base. Our 53,000 employees serve around 38.4 million customers, corporate clients, and financial institutions in over 40 countries. Our purpose is to empower people to stay a step ahead in life and business.*

> *Rabobank is a bank by and for customers, a cooperative bank, a socially-responsible bank. Our aim is to be a market leader across all financial markets in the Netherlands. We are also committed to being a leading bank in the field of food and agri worldwide.*

> *—From their website*

© Jesse Anderson 2020
J. Anderson, *Data Teams*, https://doi.org/10.1007/978-1-4842-6228-3_16

ING's Data Teams

In 2013, ING started to get serious about big data. This push came as a top-down strategy from the CEO and board of directors, who really believe in the power of data. Bas says, "Big data was not yet as important as it is nowadays. Some of the top managers recognized that they had to professionalize their data environments, and they made people responsible for that." Before the creation of the team, the people were scattered throughout the organization. ING used this opportunity to centralize them and make a true data team.

The board went even further and held strategy talks to get to the core of ING's business. Bas says, "In fact, all that we do is about handling data. And that was a strong message. The company spoke it out loud not only in the internal communications but also in the media. You can find videos of our CEO talking on YouTube and in conferences, expressing this message in a powerful way. I think that such visibility really helps to get these things into motion."

Bas was the manager of a big project to professionalize the data maturity of ING and create a production-worthy data lake. At that time, he had a team of 20 people. "And right after I heard about this initiative, I became enthusiastic and immediately asked to join the team. A lot of other people, too, either volunteered on their own initiative or were recruited by one of their colleagues. So we had a team up and running reasonably fast."

"I think that this was the situation with a lot of companies. Data is always present in a bank, but at ING, it was all spread around in old data warehouses, old Oracle databases, old COBOL mainframe applications. Our data infrastructure had grown organically over the 20 or 30 years that the bank has been gathering data. So this was one of the first serious attempts to professionalize that environment and to say, okay, we are going to build a new version of the truth for our data storage." ING was thinking through what the architecture should look like. They wanted a well-governed, well-organized, and well-modeled source of truth for the bank.

Whenever a large company starts a large project, management needs to be aware of the political implications. Bas said, "In a big organization, I think there's always politics. People want to protect the old way of working." In ING's case, team formation and rollout went reasonably smoothly. "What really helped was that immediately from the beginning, managers recognized this as a big project that was going to happen one way or on the other way. There was a lot of support from the business, from IT, and from people who were working on the old systems already."

ING is a multinational bank. One difficulty was to align the different countries because each country had different, siloed organizations. "Every country had its own data department and its own IT systems. It takes a lot of coordination to get that fragmentation in order so that everybody works with the same tools, same data models, etc. It's difficult because that means you have to align your individual units within the bank."

To achieve this coordination across countries, "the first step was to get everybody on the same tools and on the same standards." They created a reference architecture of the data lake. This aligned every country with that reference architecture and made it possible for everyone to use the same tools and data exchange protocols. "We are started talking the same language of data, basically an Esperanto that could be considered a meta-language for the data."

A big part of Bas's job was to create this alignment across the enterprise and other countries. He did this by "presenting, traveling, talking, and maybe also sometimes a bit of a push, just saying, 'This is the standard. This has been defined, and this is the way we are going to do this.' You just need some patience and some good project management basically to get these things in order."

When creating the international project, Bas found that "communication is key." They knew these international projects would be difficult. "We could trust, though, that at least the people working on data integration internationally were behind the project and were following the technology program. At Rotterdam, we were leaving them, for now, to build things in one country and then try to bring results to the other countries." To communicate effectively across international teams, they had a lot of conference calls for coordination.

ING Organizational Structure

Starting at the top, the CEO reports to the board. Underneath him is the CIO, who is responsible for all infrastructure and engineering efforts. From there, the reporting is broken up into countries, such as the Netherlands, where Bas was. Under the CIO were the various director-level managers, and under them were the department managers, which ING calls IT leads.

ING followed the agile team organization that was made popular at Spotify. Bas identifies this model as one of the bigger success factors that continues to work well for the organization. Their setup divides teams into tribes, chapters, and squads. Bas says,

"A tribe reports to a lead who is like a business manager or department manager. Chapter was the new name for a team of like-minded people. Thus, at ING, data engineering and data science were different chapters. A squad is basically a multidisciplinary team that is formed and changed along the way depending on the needs of the team."

Bas defined the role of data scientists as "to build and to train predictive models on top of data." The data engineer's role is "to run those models in a production environment goes from the data input to the data output." Engineering tasks would include "preparing the data, making sure that it's available to the model, and finally getting the results of the models and putting them into whatever is needed to use results: into an API, web application, report, or dashboard." He also points out that there is some overlap between the roles.

Bas was under the IT lead with the title of CIO team manager. After that, he became the chapter lead of data engineers. ING reorganized all the various countries' organization structures to match that of the Netherlands'. Some countries have fewer people but the same job titles, sometimes with fewer layers of management.

Part of this organizational structure includes a *meeting cadence*. This consists of a monthly meeting between the product owner and the chapter leads for each team. There the tribe leads "discuss everything that's involved in the team, including who should be on it, whether its focus and purpose are still right, the atmosphere in the team, and whether people getting along in the right way."

The squad approach allowed for even more flexibility in ratios of data engineers to data scientists. Each squad, which is made up of different tribes, generally lasts a year or more "because good squads should stay together." The makeup can change as people leave or get replaced. Other squads change because a product goes into a maintenance mode, and they need to develop something new again. "The amount and the type of work changes as it progresses."

Bas led an innovation project. "We started with one data scientist, then the team grew to three, and then it shrunk down to one again. That was apparently what the team needed, and we didn't plan it in advance, we just looked at our needs each month." This constant communication allowed them to grow and shrink the team as necessary. Management's chapter-to-chapter communication allowed people to be freed up when needed to work on a squad.

The day-to-day management of the squad is handled by the product owner. They're responsible for making sure that the squad is making progress on the project. The chapter leads are not interacting on a daily basis or responsible for day-to-day operations.

Here are some examples of how it worked. "Some teams were feature teams, building a feature on top of a Spark or Flink cluster, for instance. This, in turn, produced some functional results for an end-user, such as a predictive algorithm or a feature in the mobile app. And those teams were constructed by picking people from multiple chapters—whatever was required to build that feature. So in my case, I had several squads with only one or two high-caliber engineers. Other people came from mobile software development, from data science, or from one of the business tribes such as marketing or product development."

The ad hoc formation of squads from different tribes of data scientists and data engineers highlights the importance of multidisciplinary teams when working with data. This involves more than technical people: it brings together the product owner or the business person right alongside the team implementing the project. Bas says, "We've seen what happened when the data scientists were left to themselves, developing things out of the context of other systems. Putting data scientists and data engineers together, along with business people, allowed them to learn from each other. In these squads, the right things were happening. Creating multidisciplinary squads also allowed the data scientists to keep their focus on what they are good at, like building models that predicted patterns, etc."

Bas continues, "But data scientists could also do a little bit of the other work required in a squad: a bit of programming, documentation, or maintenance for everything that's being built in a squad. And the same goes for the other people. In that context, you can focus on your primary role 70 or 80% of your time, but you also have to take care of things that maybe are not what you were trained to do, but that need to be done to make your squad successful."

From an operational perspective, the data engineers were doing DevOps. There might also be someone in the squad from more of an operational background. Usually, the data scientists were responsible for the operational aspects of their models, such as retraining or watching for model drift.

Project Management Frameworks

ING uses Scrum for its project management framework. Bas said, "I think you can work perfectly agile with a framework such as Scrum when you're doing data science. Of course, there is some structure to building a model that you might want to follow. And you have to build in some flexibility in some cases because you don't know what you will be doing exactly the next day, but that also goes for other practices like marketing or software engineering." He recommends focusing on what's important and what you find comfortable.

Figure out where to be rigorous and where to be flexible. "When we started doing Scrum, everything was according to the book. Everything was according to the framework; it was almost like the Holy Grail; everybody had to follow the exact meetings that you should be doing when you were doing Scrum." Once they added more flexibility, "it made perfect sense to keep on using the Scrum framework."

To make sure the team is working well together, he recommends following the Scrum meetings cadence or daily stand-ups, sprint planning, and retrospectives. He goes a step further to recommend, "Don't put people in different buildings or different tables. Keeping them close together, physically helps. Also, I'm a big fan of creating a strong team culture. So team-building, going out for dinner or drinks, shared lunches, those kinds of things, those really help."

Using Data Science in Banking

Rabobank helps both themselves and their farmer customers by making farms more efficient through data science. More efficient farmers make more money and spend less, and Rabobank benefits because the farmer keeps more money in their bank and reduces their risk on loans. Rabobank's benefit also comes from an ideological perspective of helping the world's problems with hunger, reducing waste, becoming more sustainable, and improving the global food chain.

To accomplish this, Rabobank is moving into becoming a data-driven research company for the agricultural industry. This research has to be global in scope. To research on a global scale, Rabobank will have to hire many analysts and use data science at scale. The models that the data scientists build will help farmers figure out when to plant or harvest the crops.

KPIs for Data Teams

Bas has some interesting ways of setting KPIs for data teams. "I think the business results are what counts. Technical KPIs are secondary."

For the business side, he recommends determining "how many people are actually enthusiastic about the model. Give the business users ratings or surveys to see how many are using your model on a daily basis and how many are being helped." His reasoning is that, if the model isn't running or giving any business results, "it's not being used by actual customers, so it's useless."

For the technical KPIs, he recommends looking at the requirements that the model or project was supposed to do. "Is your model high performing? Is it well maintained? Is it managed? Is it monitored?"

Advice for Others

Bas recommends that "the most success comes when you have a shared purpose in one squad, in one team employing some data scientists, some data engineers, and some business people. And really have them close together again in the same structure, communicating frequently. They should talk about the same problems and the same KPIs, to celebrate the same successes."

It's also important to create realistic goals and expectations within the organization "because AI and machine learning are sometimes seen as a silver bullet. They are supposed to fix your problems yesterday. You just have to be realistic. You can't solve problems through one simple thing." Instead, you have to break a really big project into several smaller chunks that may produce several different models. By gradually creating several models, you can create a full solution.

It's important at large organizations to standardize the way you work with data, data lakes, and other patterns in technology. "I see it as part of my purpose to standardize, for example, the infrastructure that two people are working on, to standardize the way that a model is being served in production, to standardize how version control works for code. We don't want to look at five teams and see them using tools in five different ways." Those five different ways can really kill innovation and cooperation over time. The teams could spend all of their time tooling and retooling again. "You cannot reuse what has been built in other teams. So there are a lot of disadvantages to letting teams drift apart in their use of tools."

It's really important to have multidisciplinary teams. "I've seen squads deliver their highest performance when people from different backgrounds worked on a single purpose." This single-purpose gives the teams a clear priority and focus.

A company that leverages open source can hire more easily. Before I interviewed Bas, I had heard really good things about what ING was doing and the people that were there. Bas says, "I think that what we really did right is a culture of engineering in-house. I think we had a great story to tell software engineers: 'We have a passion for open source. We do agile development; you can basically pick the tools you want if you just communicate it in the right way.' I think that introducing this kind of open-source agile culture was a real good decision."

CHAPTER 17

Interview with Harvinder Atwal

About This Interview

People	Harvinder Atwal
Time period	2012–2019
Project management frameworks	Scrum, Kanban, Waterfall
Companies covered	Moneysupermarket

Background

Harvinder Atwal is the Head of Data Strategy and Advanced Analytics at Moneysupermarket. He has a master's degree in Operation Research.

> *Moneysupermarket Group PLC is an established member of the FTSE 250 index. Through our leading and trusted brands, we are committed to providing our customers with the services, tools, and products they need to grow their money. Our business model is a data-driven marketplace, providing offers to customers that they cannot get elsewhere, value to our providers, and a track record of returns to investors.*
>
> *—From their website*

© Jesse Anderson 2020
J. Anderson, *Data Teams*, https://doi.org/10.1007/978-1-4842-6228-3_17

Team Structure

Moneysupermarket has 70 people in the group data function. The data organization is headed by the Chief Data Officer (CDO). Their purview starts from the moment the data lands in a storage mechanism and continues to when an internal user accesses the data for analytics purposes.

They have a data management team consisting of representatives from data engineering, data governance, business intelligence, and data innovation.

There are two data analytics teams of different types, and both are centralized under single leaders. One, managed by Harvinder, is horizontally focused, while the other is vertically focused. The vertically focused team works on all of a brand's products. For Moneysupermarket, those verticals are utilities, telecoms, money products, and travel supermarket. Harvinder pointed out that it's unusual to find the analytics department under the CDO. Usually, "they're in IT, although you may find some central analytics teams embedded within some of the business units."

Harvinder's definition of a data scientist is "someone who undertakes advanced analytics." They primarily create models that benefit their customers and internal processes. "They're also involved quite heavily with the product teams. So we do a lot of optimization testing of the website, and some AB testing. It could be done in batch form; increasingly, we are also looking at streaming pipelines."

The company's definition of data engineers has changed over time. They used to have more ETL developers. Harvinder described their role. "They helped to get the data into a format for the data analysts and data scientists to use." They used to take some of the machine learning models and recode them.

Operationally, the team is following a DevOps model. "If you build it," says Harvinder, "you own it and run it. Data operations are really a central function that sits under data management and looks primarily after our databases and servers. So we have decreased our reliance on operations for the data analytics function."

Moneysupermarket is moving to the squad-based approach, similar to Spotify. This will allow more autonomous teams that will have all of the skills to accomplish whatever is necessary.

Harvinder says, "We have realized that we don't need all the data scientists we have, and that actually they create imbalanced teams. To create a nice balanced team, we're looking at getting people in with a sort of software engineering expertise. We also want people with data analyst backgrounds so that we can cover a wider spectrum of skills and cross-train each other. Even if you get the right skills within one team, you can still create silos out of individuals. So one of the other things we're keen on is to cross-train people. They'll never be specialists in each other's areas, but as long as they know enough that they can help out other team members and keep the flow of work going, that's better than having a team of specialists."

Removing Barriers and Friction

Moneysupermarket started out working with data in a more traditional sense. The teams of data scientists and data analysts were there for more decision support. But the company realized that the needs of the organization and the techniques it used were going to change over time. Originally, IT was using data warehouses and focused on governance and control. But management wanted to create more advanced machine learning models. The consumers of data wanted more access, freedom, and flexibility around their use of data. The platform actually hindered the data teams from being able to use the data.

To accomplish these changes, Moneysupermarket would have to remove all these barriers and friction. The changes were both organizational and technical. They realized that, as Harvinder says, "It's not as if we're going to buy a particular tool or a particular platform to solve all our problems, or that we just need to be organized better to solve all our problems. There are many things to change."

From an organizational point of view, the company had the "wrong kind of work for the wrong people, which was wasting people's time." They had to change "where teams sat, who they reported to, and their responsibilities."

From a technical point of view, they had to change the tools and platforms they were using. But they couldn't remove all of the governance and controls on data because they were storing PII (personally identifiable information) and had to ensure it wouldn't be misused. The team had to create a platform that was still secure without hindering use.

As part of these organizational and technical changes, the data teams organically started to gravitate to a DataOps model. Harvinder defines their DataOps as "a combination of DevOps, lean thinking, and a sort of agile methodology."

Their move to DataOps is a constant journey because there are always things to improve. In particular, Harvinder is focused on improving "data management, data governance, and the monitoring of our data flow."

The main improvement they have discovered when using DataOps at Moneysupermarket has been increased speed: they create data products faster and using them faster. "From the ability to speed up the way you can use data, DevOps definitely helps a lot in analytics." The speed improvements are helping out several departments in the organization, such as marketing.

When it creates functional teams, the business has to triage priorities with several different departments, teams, or even CxOs. Each team has its own backlog, and their priorities may not match the business' priorities. This leads to prioritization requests that have to constantly be escalated to senior management. Each squad or team should have every skill on it that is necessary to accomplish the task.

Having all of the data teams under a single C-level executive also reduces friction. Earlier, some teams at Moneysupermarket were under the CDO, while others were under IT. This made visibility into changes and updates much more difficult. Consolidating under a single executive enabled much better communication. Harvinder says, "We have more opportunities to talk to them about what we want from them and how we use the data, so it has been incredibly beneficial." The reorganized improved goal setting for the data teams. "It's much easier to make priority calls so that we're not working on completely different objectives."

Part of reducing friction requires aligning the data teams with the business. "We don't want an analytics or data science team to just go off and do our own thing, experiment on that, and then come back and find that people aren't interested in what they produced because there are other objectives." Instead, they start by aligning with the business first and figure out how analytics can further their objects.

This planning starts with the road map. The data teams sit down with their stakeholders to understand their road map and how they can contribute. By moving to squads, "we're very much aligned on objectives. We find that we're much more collaborative. It's much easier to persuade the business that we've got a really good idea that could help contribute. They can see how it will help them and benefit them."

Good alignment is necessary to keep the whole system functioning. Without it, there is a good chance the team's work doesn't make it into production, and this presents a "real risk of that work is wasted." When things are wasted, the data scientists and data engineers don't see their efforts making a difference. When people see their time and efforts wasted on a lack of alignment, they will go to other companies. That's "another reason why alignment is really important for us."

Not everything that teams do will appear on the business' road map. "There still has to be some room for innovation, as well as for going back and clearing out the technical debt that we've built up, and setting up new technologies."

Project Management Frameworks

Moneysupermarket uses Scrum, Kanban, and Waterfall project management frameworks. Primarily, the teams use Scrum, but the management leaves the decision up to the individual teams. "We're not too rigid about enforcing certain ways of working; we find that different ways work naturally better for others. We're a little bit more rigid around things like enforcing certain types of best practices. For instance, we do have a software development life cycle."

Some teams may use Waterfall because there is a specific end date. For example, this could be when a license for a legacy technology is about to expire. The team needs to plan to hit that deadline.

Data science and analytics teams have found that Kanban works better for them. "Because of the exploratory nature of our work, we don't always know when we're going to have an acceptable result or one that we're happy with." In data science or analytics, you may not have "a shippable product at the end of that increment."

The amount of process also depends on whether the result will go into production for repeated use or is a one-off insight for an internal team. If something is going in production or might be dealing with sensitive data, they put a software engineering process and rigor around it.

Some data scientists chafe at software engineering processes. They may feel that it is too much of a challenge or that the processes are too heavyweight for what they're trying to do. So Moneysupermarket trained the lead data scientist in each team on software engineering methodology and showed them how to use it in ways that are effective for their team. The lead data scientists, in turn, trained their teams to use it. It was really up to the lead data scientist to enforce software engineering best practices in the team. The data scientists learned that "if you do it right, it's there to save you from making mistakes, which would just waste your time. So, the process is there to help you, not hinder you."

Team KPIs and Objectives

All data teams' KPIs are aligned with the organizational and technical objectives. The technical KPIs monitor things such as data quality, data completeness, SLAs, and timeliness of data.

For the organizational KPIs, the actual metrics are not about their output, but about the outcome. "They have a lot of freedom in how they achieve those outcomes, but the only way they're going to achieve them is if they collaborate with their stakeholders."

This requires good communication with the business. Otherwise, they could create incredible models, insights, or data products that go nowhere.

The organizational objectives are set once a year. Moneysupermarket's bonus structure is more like a traditional financial company, in that year-end bonuses are the objective people want. "We're a publicly invested company, so it's a reasonable amount of pressure to hit your target and make shareholders happy." Harvinder monitors the team's progress on their organizational objectives throughout the year. "Every quarter, I'll get the team together. We'll see whether we're on track. We'll also do retrospectives to see what we can do within the team to improve our performance."

Changes in Data Teams

Machine learning presents new challenges that regular analytics or software engineering doesn't have to deal with. Harvinder compares it to a calculator. If you were to write a calculator, you'd always want it to create the correct calculations forever. With machine learning, however, the calculations are always changing. Once a model has been created, there are issues such as changes in input data that create model drift, and the model needs to be constantly monitored for that drift. If the model does drift, it won't give accurate predictions or results.

This leads to the question of who is responsible for watching for the drift. That person would need to decide what the best course of action is. Should the model be retrained? Is it time for a completely different model or approach? Harvinder thinks it could be one of the responsibilities of a machine learning engineer.

Harvinder sees data engineers "as being really essential to what we do, rather than being a separate function that just takes orders from us, delivers the data set when we ask for it. They must be proactive." Having the right data engineers can make the rest of the organization more productive and efficient. "I think we probably need fewer data scientists overall, and we probably need more support from our data engineers because we already have a lot more data than we can integrate, along with new data sources all the time."

Organizations need to make sure that everyone who contributes on data teams is given credit for what they've done. Usually, the data scientists get the majority of the credit because they're the most customer-facing part of the team. "But it's actually a success shared by everyone else too. I think that being part of a DataOps team lets them make more of a contribution, and get more recognition for what they do as well." It's up to management to make sure that stakeholders know who contributed. "The person who delivers the good news shouldn't get all the glory."

Advice to Others

When organizations are starting out with data science, they will hire "some really exciting people and then just tell them to go off, find something interesting to do, and provide some insights." The data scientists will go and find those insights, "but what they're coming up with is not part of the roadmap or objectives of the stakeholders." The data scientists will then find it "incredibly demotivating" that none of their models or insights get used or move into production. Harvinder recommends not seeking out tasks that are "particularly sexy or bleeding edge, but to get buy-in and implement something really amazing."

New data scientists can have really unrealistic expectations of what they're going to be doing once they're hired. They think they'll be spending all day creating machine learning models and tuning the hyperparameters. A data scientist is "paid to use data to drive decisions." This focus on models instead of value creation is fueled by conferences, articles, and competitive data science competitions. "But I'd rather have a good model that actually gets used in a product."

For larger organizations to be successful with data teams, Harvinder recommends getting the basic foundation solid. The organization needs to create a clear data strategy. That means having an answer to "how are you going to acquire data, how are you going to store it, and how you get the data governance, data management in place." The organization also needs alignment its strategy so that everyone understands the end goal and how each person will individually contribute to the goal. Management needs to make sure there aren't any gaps in the required people or data teams. Management will need to make sure the people have all of the required skills. Without solid foundations, any data strategy or data project won't get far.

Smaller organizations tend to hire a solitary data scientist who is expected to fulfill all of the responsibilities of each data team. This data scientist is often an inexperienced person. Instead, startups should make sure they're "recruiting the right individual for that role."

Then there is the data itself and the amount of data. Some small companies "just don't have enough data to do anything interesting with it." The startup needs to be realistic about how much data they will start out with, and how long it will take to build up enough data for reliable analytics. Without enough data, it can be really difficult to baseline or actually track the improvements made by a data scientist.

On the data engineering side, Harvinder recommends emphasizing "data analysts, data stewards, data lineage tracking, and metadata." At Moneysupermarket, their data comes from the website and is relatively clean JSON. Even then, they can have challenges around data quality or data lineage. Harvinder would have invested "more time earlier in building out the team for software engineering, and making sure to have software development skills along with the data scientists and data analysts."

CHAPTER 18

Interview with a Large British Telecommunications Company

About This Interview

People	Anonymous Architect
Time period	2013–2019
Project management frameworks	Waterfall and Agile
Companies covered	A Large British Telecommunications Company

Background

This interview is with a person from a large British telecommunications company (TC). In order to publish the interview, we have to keep both the company and the people anonymous.

The anonymous person (P1) is the Chief Data Architect. He's been with the company for two decades, serving as their architect as the company has transitioned from one technology to another over the years. He's responsible for data management all the way from traditional databases through big data and machine learning architectures. He has an MBA.

© Jesse Anderson 2020
J. Anderson, *Data Teams*, https://doi.org/10.1007/978-1-4842-6228-3_18

Starting the Initiative

From a data engineering perspective, organizations should maintain a lot of flexibility. You want to avoid creating an end state or final look too completely. The organization's view of the final state will change for technical or business reasons. P1 says, "it feels like we're in a permanent state of evolution, in that we kind of know what we've launched and what we have in production, but we're not quite sure what the next thing will be." The architecture needs to support the gradual changes and technical innovations that data scientists need to do their job. It's the architect's job to anticipate those changes.

Creating an architecture requires a delicate balancing act. It requires the working teams using the products to understand what they really need. Once there is a good understanding of the use cases and ideal workflow, P1 has to look at the existing technologies for any gaps. These could be gaps in technical features or in terms of scale and time to complete the processing. "The job involves trying a piece out, iterating, and then growing to where you get success."

When data teams are just starting out, the trickiest part is to figure out how to work together. P1 recommends starting out with just two projects. That will "build shared understanding with the teams that have got to do the building, with teams that have got to do the consuming, and perhaps the teams that are supplying data to you. You can get enough capability so that everybody understands their roles and responsibilities." From there, things will start evolving.

Working with the Business

TC has always had a strong research and development background. Thus, they started out with an applied research team, which evolved from a deep background in quantitative math and improved their coding skills. This allowed TC to create a centralized data science capacity, where others in the company can find highly specialized skills in telecommunications.

From there, an increasing number of business units have two to six data scientists embedded in the business unit. These business units include consumer marketing, network operations, and portfolio planning. The job of data engineering and architecture is to provide those teams with the tools they need now and in the future.

One example of the evolution in their data analysis was aimed at optimizing their call center capabilities. Traditionally they had static expert systems that simply looked for keywords from the caller. After the data scientists created more sophisticated models, the call times went from 40 minutes to 5–10 minutes. There was an engineering difficulty in connecting all of the data with the various systems that needed to be integrated, and an organizational difficulty around taking ideas from the business units and coordinating across several different departments. The call center project helped save money and improved customer service through shorter call times.

Data Scientists and Data Engineers

For P1, the biggest gap is how the data engineers should work with the data scientists. With large enterprises, data engineering becomes difficult because of the sheer number of systems that need to be integrated. To get a 360-degree view of the customer, the engineers must combine several different systems that weren't designed to integrate with one another. Added to this architectural difficulty is the buy-vs.-build decision. Can the company buy something off the shelf that will do what they need, vastly accelerating the project with minimal customization? With the added complexity of distributed systems, this decision becomes even more difficult.

Supporting Models for the Enterprise

Modern machine learning models tend to be used by enterprises much longer than older forms of data products. Enterprises have to think in cycles of longer than ten years, so their models must have the same longevity. Most enterprises have shorter-term processes in place for software engineering and outside vendor relationships. Now enterprises have to create long-term plans for models. A large risk is inherent in deploying a model long term in a crucial function, such as network automation.

Therefore, P1 says that TC is "better off with the solution we've built ourselves because we can build it carefully and tune it more deeply to our particular customer base and our product set. We can build a better, more accurate model than an outsider." They've also been able to productionize models much faster and more extensibly by creating their own serving systems. These homegrown systems also enable their data scientists to build models by making sure good data products are exposed not just to serve model building but for discovery and exploration too.

TC also had to face the question of who should support each model over the long term. TC's legacy operational support teams lack the skills for such support. The task may fall to a data engineer to be taught how the model works and how it should be working. They'd be assigned a tolerance for the confidence intervals and would make sure all scoring falls within that tolerance. If the model failed to do so, the data engineers have instructions on how to retrain the model. Finally, they're given a plan for a fail-safe state, or what to do in case everything blows up and isn't working.

Moving from Proof of Concept to Production

There is a danger of Scrum teams or squads becoming siloed. Teams can get overly focused on their project and fail to look at what the rest of the organization needs or is doing. As an organization moves from well-defined and segmented projects to serving up data as a service or product, this project focus will become a problem.

For more proof-of-concept and speculative work, TC created their big data platform and data products with "a very simple standardized service offering and a very clear definition. It was very much the classic Ford model. You can have anything you like on this platform as long as it's one of these, and it comes with one half a terabyte of storage, one cube with preemption enabled, and your job may be killed." This big data platform was exposed without a chargeback model, and that helped dictate the rules of its usage.

This model worked well for onboarding teams quickly. But it was a big change for the infrastructure team because they were used to spinning up VMs and never dealing with them again. Running a platform or service meant the infrastructure team had to become aware of what was running and monitor those processes. Knowing what was running was crucial because "we are faced with what we refer to as 'noisy nightmares,' tenants on the platform who are not fully educated, who are suboptimal in their skills, and who are causing problems throughout the production." So it required a mindset change to run a service platform.

The trade-off for proof-of-concept work is inhibiting innovation by putting too many engineering rules or SLAs in place. P1 said, "One of the great things I think we've managed to do is dramatically lower the bar to doing a fairly speculative data discovery or insight, one where a small team wants to understand a problem or do a one-off analysis of where should we invest. This is in contrast with a large program, which has fairly clear requirements, funding, and a very clear line of sight to its benefits. But we have to accommodate both of those on the same kind of infrastructure, which in itself presents challenges."

As a product moves out of the proof-of-concept phase, the enterprise has to think about how to productionize the infrastructure. For TC, they "naively thought we could get everything on one cluster, but that clearly wasn't going to be." Now they use dedicated clusters for most use cases. That way, TC prevents someone from "coming along and writing some crazy job that would destabilize everything." Creating this production cluster included creating its SLA. Those SLAs need to meet both the business and technical requirements of the use case. By launching multiple clusters, a large organization can use the pet and cattle analogy when dealing with operational issues. A single monolithic cluster means you have to maintain the cluster as a pet. With multiple clusters, you can deal with them like cattle.

Creating Enterprise Infrastructure and Operations

The first stage of TC's big data journey was a multitenant environment. Their initial goal was to create a place to democratize data and use cases. This allowed the business to start creating and running analytics. The problems started when the business started to put these analytics and workloads into production. The business started relying on these "production" data flows and expected the data engineering and operations teams to support them with SLAs. But the team was still creating the processes and workflows to deliver on an SLA.

This disjunction between the team's capabilities and the business' expectations led TC to put more effort into getting the business requirements together: a clear definition of the data product and the business requirements for that data product. TC set up a process that lets the business managers define their business requirements and their desired data products in a clear way. From there, they run a proof of concept, define technical requirements, and create a minimum viable product (MVP). This process includes regular touchpoints with the business that allows the business to decide whether the functionality meets, exceeds, or is below expectations. This iterative process keeps evolving until the business can see that the data product meets their needs.

A key subtlety for both business and technical people to understand is how data products differ from typical programming artifacts such as modules, API endpoints, and software libraries. To help the business understand this difference, TC has instituted a centralized data engineering team along with separate data engineering teams within business units. The centralized data engineering team reviews proposals from each business unit team to make sure it follows best practices. The process of validating the business unit's proposal is iterative. It isn't just a one-time submission to a committee. Instead, the business unit has the full cooperation and brainpower of the centralized team to iterate on the proposal until both sides are happy with it.

TC was always very rich in data. However, they would keep only less than 1 percent and throw out the other 99 percent. This was because the team was focused only on one use case of the data, which required only that 1 percent. In this use case, the data was used only to handle failures. Because the other 99 percent of data was thrown away, the later use cases could only look at failures instead of the way that the missing data could be used.

So TC started to capture all data so that they could process all types of use cases. This broadened the applicability of the data from just one specific and technical use case to be usable by the rest of the business.

Project Management Frameworks

The majority of TC's IT projects use Waterfall methodology. The projects mostly operate on three-month delivery cycles. Waterfall is used because it allows big-ticket items to be released all at once and for all of the various teams to coordinate a much bigger release. The teams using the data team's platforms are more focused on predictable timelines than agility. The tenants of the platform want predictable performance, with a low desire for new or updated components.

There is an effort underway now to start using agile methodologies for the data teams. In order to accommodate tenants who want few to no changes, the company is transferring their projects to a separate, dedicated infrastructure. Other teams that want faster and more agile approaches stay on the current infrastructure.

Advice for Others

TC realizes the importance of people in being successful with data projects. They try to hire people who know what they're doing. There are a lot of people who say they can create big data systems but they've never done the work before. Their knowledge is all theoretical and it's crucial to have people with real-life experience. When people lack practical knowledge, they're building on a foundation they may not fully understand and can't understand the implications of their decisions on architecture or operations. Fundamental mistakes like these will be incredibly costly in the long run.

It is critical for an organization to get the right support and investment from the business before starting on a big data journey. The business has to invest in the right data teams and personnel from the very beginning. If a single person or a small group is trying to do everything, the improvements or ROI will be minimal. The organization won't be able to realize its goals in leveraging data.

P1 recommends that management understand and internalize the changing landscape of technology for distributed systems. A technology's usable life is getting shorter and shorter. In a short time, P1 has seen several technologies come and be supplanted by newer and better technologies. P1 says, "The impressions I get are that planning time horizons are becoming shorter and shorter, while the business just wants to go faster and faster. It wants to get stuff done right now."

The speed with which businesses demand change is forcing data teams to change. Data teams have to think about how to get an idea off the ground quickly and then productionize it so that the model is scalable. There may be a team that wants to do something within a single business unit, but the same idea or workflow could be broadly applied to other parts of the organization. P1 sums it up by saying, "Start small, go fast, deploy quickly."

CHAPTER 19

Interview with Mikio Braun

About This Interview

People	Dr. Mikio Braun
Time period	2015–2019
Project management frameworks	Scrum
Companies covered	Zalando

Background

Dr. Mikio Braun is a Principal Researcher at Zalando. He has a master's degree in Computer Science and a Doctorate in Computer Science and Machine Learning.

> *About Zalando: "As Europe's leading online fashion platform we deliver to customers in 17 countries. In our fashion store, they can find a wide assortment from more than 2,500 brands."*
>
> *—From their website*

© Jesse Anderson 2020
J. Anderson, *Data Teams*, https://doi.org/10.1007/978-1-4842-6228-3_19

Organizational Structure

Data is crucial to an ecommerce company. Zalando started out its data science journey with a centralized team called *data intelligence*. They were mostly working on creating small prototypes.

In 2015, there were about 30 data scientists. Zalando's VP of Engineering, Eric Bowman, spearheaded a *radical agility* project to transform the company organizationally and technically.[1] The company decided to break up the centralized team and distribute the data scientists across new, cross-functional teams aligned around products. This brought back-end engineers and data scientists together into the same teams, along with the product and business side. Each functional group was organized into *guilds* so similar functions could get together and exchange learnings.

Each team was responsible for deploying their own systems, choosing the technologies to run, and all operations for it. They were worried that a centralized cluster would be too difficult to get right for all teams and use cases. Upper management thought a decentralized approach would allow the greatest flexibility for each team to make its own decisions.

Decentralizing the teams and decision-making had an unintended consequence. The proliferation of independently developed and deployed systems made it hard to prepare and maintain the systems for production-ready data projects.

In 2017, Zalando started forming data engineering teams focused on infrastructure. They improved the data infrastructure supporting the data scientists and other teams. This included centralizing the machine learning infrastructure. Since there was already internal experience with creating platforms, the team was able to leverage their experience to choose what worked and discard what didn't.

The common infrastructure started improving operations and the sharing of best practices. Having centralized infrastructure allowed the team to better handle software that was hard to set up, by capturing its configuration and automating its setup. Adding automation allowed them to speed up all parts of the process from development to production. There wasn't a mandate to move to the new infrastructure and adoption was slower and mostly focused on new teams.

---

[1]Eric did an in-depth interview on the organizational changes he made at Zalando, www.mckinsey.com/business-functions/organization/our-insights/the-journey-to-an-agile-organization-at-zalando.

As new teams started doing machine learning, they were able to leverage the learning and centralized infrastructure. Zalando wanted to prevent new teams from having to spend years having to relearn the lessons that previous teams had already experienced and solved. Through continued work, Zalando improved the user experience of taking a model from development into production.

Zalando structures its teams around research, product, and development areas. Mikio was the team lead for the search and recommendation team. After that, he was promoted to Principal Researcher, where he spent his time looking at machine learning across Zalando.

Socializing the Value of Doing Machine Learning

One of the biggest challenges with any machine learning project starts with the data. Acquiring the data involves working with many different parts of the company. The teams who were producing the data at Zalando, usually front-end or back-end engineers, weren't the ones versed in machine learning or creating analytics on the data.

Zalando generates extensive clickstream data that needs to be analyzed. This clickstream data comes from the front-end teams. In the beginning, the front-end teams didn't understand how valuable the data they were creating was for the company and other teams. To the front-end teams, the data was just a method for tracking and debugging issues across the systems. They weren't aware that other teams could take the same data and use it in a completely different way. By not fully leveraging their data, they were leaving business value on the table.

To convey to the front-end engineers the value of their data, the company had to change the engineers' culture and value perception of data: to explain that their data could be used for several different purposes. This perception and appreciation of data are really what separates the data engineers from the front-end and back-end developers.

It's important to create realistic expectations of new technologies and trends. Trends such as machine learning are especially vulnerable to the hype from the media and vendors. Zalando started to create realistic expectations by ensuring the product managers and product people understood what machine learning is. They accomplished this initiative through their training programs. The training programs were joint effort support by an enthusiastic group of individuals that included Mikio, an AI enablement project manager, and the internal training team.

"I actually figured out that the problem that needs to be solved before [we can create useful data products] is that everybody who's involved in them understands better how to do data science-related projects," Mikio said. Both the product people and the technical team leads need to know the differences created by the new machine learning techniques, and that "you would do a pure engineering project differently from one that involves machine learning."

Upper management also needed to understand the value of machine learning. Mikio together with the internal training team provided this background as half-day workshops aimed at the company's SVPs and VPs. This helped to separate the hype from reality. The workshops helped them understand the difference between machine learning and other analytics, and the potential strategic impact of machine learning on the business. The training stressed that the right data had to be acquired first before the data science teams could leverage it. They talked about the need for long-term historical data to get the best possible results from the machine learning algorithms. Since you can't go back and re-create data, the workshops stressed the need to think about the future uses of data when deciding whether or not to save the data for the long term.

To make sure the knowledge wasn't confined to upper management, there was a separate workshop aimed at nontechnical people in the company. Their focus was to consider machine learning at a higher level and highlight the data-driven ways that problems could be solved. In the end, over 300 people participated in the workshops. This helped the nontechnical people incorporate data and machine learning into their strategic plans.

There were also more technical sessions aimed at back-end engineers. This allowed back-end engineers who were interested in machine learning to start deepening their understanding of it. The sessions were followed by more intensive training where the back-end engineers could learn the basics of machine learning. They could then start to work together with the data science teams on machine learning projects. By offering several gradual steps of increasing complexity, the company could serve all their employees while weeding out the cursory learners from those who would commit the time and effort to learn.

Definitions of Job Titles

At Zalando, there is an entire job family called the *research job family*, containing both data scientists and research engineers. The job description for data scientists follows the definition set out in this book. Their definition of a research engineer is akin to a machine learning engineer, who tends to be much better at coding than their data scientist counterparts.

There is another job family for software engineers. Data engineers are part of this job family, along with front-end engineers and back-end engineers. Data engineers are specialized in big data technologies and infrastructure. This allows data engineers to support the data scientists in their efforts. Some back-end engineers also have a strong background in big data and an interest in data.

There isn't a specific operations team or operations title because Zalando does DevOps. The teams themselves are responsible for deploying and supporting their own code. An SRE team does the first level of support, and the teams provide the second level of support.

Zalando has a career path for individual contributors that goes up to the equivalent of a Vice President level. The Principal Engineers aren't people managers, but instead, curate the technical expertise and provide technical leadership in the company. Principals are brought into bigger projects to validate the architecture and verify the technical solutions. By having a high-level career path, engineers aren't forced to go into people management to feel like they're advancing in their careers.

Reducing Friction

In 2015, Zalando was experiencing a great deal of friction organizationally and technically. They were spending too much time trying to negotiate resources and time across teams. For example, the data science teams consistently needed the engineering team's help. Friction was compounded by a monolithic code base running in a single datacenter.

Better access to a back-end engineer's expertise solved some of the difficulties for the data science teams. Data scientists were good at creating small prototypes. However, there is a significant difference between a prototype and production quality code or models. The data scientists could create a prototype but needed another team to harden the code to production quality. The other production team had their own priorities and tasks. This usually meant that all of the production team's time and resources were already spoken for. The data scientist and production teams would have to negotiate and triage the requests for their resources.

Zalando realized that this limitation, exacerbated by its highly top-down structure, would inhibit growth. The resulting friction of negotiation and constant triage brought about an inability to scale the organization.

Zalando started combining people from different job families into a single team to reduce friction. This organizational structure is close to the DataOps definition set out in the book. Each team would contain both data scientists and back-end engineers. This allowed the team to create the whole pipeline.

The team would work with the product people to understand the requirements, create a prototype, and then create the final production system. The back-end engineers would support the data scientists on the coding and technology choices. The data scientists would create models or analytics. Now, the team can deliver the entire product without relying on another team's resources or timelines.

The original choice for combining the teams came from VP of Engineering, Eric Bowman, who decided first on a technical switch to microservices and the cloud. As part of the switch, he reorganized the data science and engineering teams into autonomous end-to-end teams. Each of these teams would just work on one part of the service.

The congealing of employees into end-to-end teams mostly removed the need to reach across the organization in order to accomplish a goal. Now, the data scientists had direct access to a back-end engineer and didn't have to try to reach across the organizational boundaries to get a back-end engineer's time. The end-to-end teams removed the bottlenecks and improved the velocity and growth.

Project Management Frameworks

Most teams at Zalando use a form of Scrum methodology. The Scrum framework puts a greater emphasis on regular check-ins like daily stand-ups and weekly meetings. "I think that's actually pretty similar to what you have in real research. So when I was at the university and supervising Ph.D. students, they also would just meet every week and look at what's happened in between to talk about where to go," Mikio observed.

Scrum can be difficult to use with data science teams because it's "very hard for data science to commit to that [results-oriented framework] because so much is unknown," Mikio says. To make the framework work better, they "redefine what the outcome at the end of the sprint is, and you also make it more a kind of research or easier to use as research." For example, a sprint's outcome could be defined as "determine whether a new algorithm is better than the baseline" instead of "deploy a new model based on this algorithm." These research-focused outcomes allowed data scientists to commit to doing something in two-week sprints.

This interpretation of Scrum allowed Zalando to have frequent check-ins to verify the data scientist wasn't stuck or no longer believed the algorithm to be viable. As the data scientists worked, they could better articulate the possible outcomes of their sprint's work. It's during these check-ins that they discuss what could be done differently. "But at some point, you also need to think about whether you want to continue at all."

KPIs

Zalando's KPIs for data teams revolve around the product they're responsible for. For example, the recommendation team's KPIs focus on purchases of products that are promoted by the team's recommendations. Other teams would have similar KPIs that are based on the data product or model they are creating and the value created for the business.

Improving Data Scientists' Engineering Skills

Going from prototypes to production code required several changes in how the data scientists thought and acted.

It started with improving the software engineering discipline around coding practices. Teams needed to make sure the code followed software engineering best practices in order to be maintainable over the long term.

At the start, data scientists indulged in certain practices during the exploration phases that continued on during the release to production. The data scientists were deploying code changes directly to production branches and had to be taught that this isn't a good software engineering practice. The teams also required code reviews. Instead of just dictating these changes, they made sure the data scientists understood the reasons these are accepted best practices in software development.

The culture change extended into some the finer and more stylistic aspects of software engineering. Zalando started to educate the data scientists on how code should be structured and why. It helped the data scientists think about the long-term impacts of code quality.

To verify that code was production-worthy, Zalando put several processes in place. They made sure all code was checked into source control and had been code reviewed. As part of the review process, the code is checked for performance issues and adequate unit tests.

The entire process is set up to mentor the data scientists. Zalando understands that change is a process that takes time. Even software engineers with four-year degrees need to learn how to write production-worthy and maintainable code. "I think it's a mentoring process because, when the data scientist sees the good layout, the data engineer explains why they do that and what they do. Data scientists are smart enough to start doing that as long as they have an interest in the code side," Mikio suggested.

Integrating Business and Data Teams

Instead of having a business side and a technical side, Zalando organizes their teams around an area, for example, a single integrated team that handles the entire area of merchandise buying. The team includes the business people making the purchases, the marketing efforts, and the data teams handling the technical aspects. By having fully integrated teams, the business gets exactly what they want and is able to dictate the priorities.

This wasn't always the case at Zalando. There used to be a separation between the technical and business departments. After a reorganization, both types of teams were brought together.

Because the data teams are part of the business units, the technical people aren't centralized. A lack of centralization for data teams can make it difficult to pass on best practices and maintain technical quality standards. Zalando started using a new organizational structure called *guilds* to keep data teams communicating and working together. These guilds include weekly talks to foster the exchange of information and best practices.

The Differences Between European and US Companies

Given that Zalando is a European company, I asked Mikio about the cultural, technical, and organizational differences between the two regions.

Mikio brought out the point that European companies take a different approach to agile implementations. Instead of having separations between or hand-offs between departments, European companies often have a single team combining different job titles. They're more agreeable to creating teams aligned to a product rather than title-based or function-based teams.

Teams in Europe are often more culturally diverse than companies in the US. In the US, teams are made up of 1–3 different cultures, whereas a European team can have 5–7 different cultures. Without monocultures, data teams can escape being trapped in a single understanding of ideas. The complementary contributions of cultural differences help teams be more productive and get stuck less often.

The General Data Protection Regulation (GDPR) is another big difference between European and US companies.[2] The GDPR directly affects data teams because they are responsible for data storage and processing. Mikio recommends thinking about customer data and regulations from the very beginning. The teams can think of technical strategies to comply with the GDPR, such as anonymizing or encrypting data. Teams need to have a plan for deleting an individual's data. Addressing regulations too late in the design process can be painful because it will require code rewrites and perhaps even architectural changes.

Advice to Others

It is important to "have a clear idea of what you want to deliver," Mikio recommended. The clear idea includes the project implementation and clear KPIs "so they have good KPIs that you can optimize against." These clear goals and measurements have created a virtuous cycle at Zalando that made them comfortable with continued investment in machine learning. "I think the one thing that has always been true for Zalando is that you have a lot of support for putting resources into machine learning."

At the foundation of every data team is the data itself. "I think that's why it's super important to start building good data." When a company is just starting out with data, Mikio recommends not trying to collect all of the data at once. Instead, the data teams should focus on collecting the "data in a way that has quite high quality and that can be found well." The data team's initial goal should be to have the right data, well formatted, and findable, so that "whenever you want to try out a new idea on some data that you already collected, it should be really easy to do so." This ease of analyzing or processing

---

[2]GDPR is an EU-wide directive regulating data protection and privacy. It has the force of law and can enforce substantial financial penalties. Some US states and other countries are putting in GDPR-style regulations. This difference will start to become the norm for data teams.

data keeps a team from creating a "one-off hack on the existing infrastructure" for development purposes, followed by "the one we scale at the end" by rewriting it for production purposes.

Some best practices need to be in place before a company splits up centralized data teams into product teams. For Zalando, it was critical to establish a data engineering culture before splitting into different teams. "So I think that helped to set the tone" for teams to continue the good data engineering practices they were doing before the split.

Recently, Zalando started to ask the question, "what stuff built by one team could be useful for some other team?" This inquiry leads to the creation of general-purpose code and shared libraries, which Mikio wishes they had started earlier. "We take what works well for one team and then organically build it up and make it useful for other teams" so each team isn't creating their own version of the same thing.

Such redundancy isn't relegated to code, but extends to data too. Different teams in different areas were processing the same data in a similar or slightly different way. The teams had access only to the same base data. There was no way to create a modified dataset that exposed an enriched or aggregated dataset to be shared across teams. As a result, each team created its own representation for the enriched or aggregated data. There wasn't even a mechanism or process for each team to realize they were re-creating the same data each time.

Zalando's teams now really understood their goals and KPIs. They have good metrics to know how their products were measuring up against their KPIs. A focus on improving their KPIs fostered improvements and encouraged teams to try new approaches. The KPIs created "proxies for evaluating new models offline" to quickly and easily see whether a new approach was better than the current approach. They could "focus on making improvements on the machine learning side." Instead of having to think "I have to find the data, I have to plug everything together," there were already frameworks for evaluation in place that made things easy and automatic. "It becomes really easy to try a new idea and then very quickly see whether it's actually better than the thing that they had before." This created a virtuous cycle where the barrier to experimentation was incredibly low and more experiments could be run to find the optimal solution.

Index

A

Area Under Curve (AUC), 235
Artificial intelligence (AI), 29
Assigning tasks
    collaboration, 138
    data engineering team, 138
    data science team, 138
    effects, wrong team, 139, 140
    level of triage, 137
    operation team, 138
    rectifying problems, 139

B

BackTape, 244, 245
Big data projects, 95, 215, 218
    business complaints, 22
    cant definition, 4
    complexity, 6
    data engineering team, 13, 14
    data pipelines/
        data products, 7
    data science team, 12, 13
    definition, 5
    distributed system, 6
    failure, 21
    general answers, 23, 24
    iterative approach, 185
        crawl level, 186
        evaluation, 185

management, 5
misconceptions
    business intelligence, 8
    data warehousing, 9
    operations, 9
    software engineering, 9, 10
missing team, 17, 22, 23
new technologies, 22
one-offs, 25
operations, 15, 16
planning/starting, 177
    data spread, 178
    level of complexity, 177
    understanding of business, 179
presteps, 173
    big data problem, 176
    business need, 175
    data problems, 174, 175
    execution strategy, 176, 177
programming skills/missing tool, 25
scale, data science, 24, 25
small organizations, 17
starting project
    actual code, 183
    compute clusters, 183, 184
    consultant, 179, 181
    goals/metrics, 184
    technology choices, 182, 183
    third-party vendor, 179
    training, 180

© Jesse Anderson 2020
J. Anderson, *Data Teams*, https://doi.org/10.1007/978-1-4842-6228-3

Big data projects (*cont.*)
    success, 20
    success/failure, 11
    technical reasons, 4
    technologies, 5
    3 Vs, 4
    underperforming/failing teams, 21
    wrangling, 25
BizOps, 112
Bottom-up approach, 235, 236
Brick-and-mortar retailer, 237
Business goals, 221
Business intelligence, 8
Business interaction
    areas, 110
    BizOps, 112
    business intelligence/
        data analyst, 132
    data-augmented organization, 127
    data products, 109
    data strategies, 123, 124
    data warehousing/DBA staff, 130, 131
    domain knowledge, 112
    executive-level attention, 127
    funding/resources (*see* Funding/
        resources)
    individual technical
        contributors, 111
    insufficient/ineffective, 118
    KPIs (*see* Key Performance
        Indicators (KPIs))
    managing/creating realistic
        goals, 125
    medical insurance domain
        knowledge, 113
    middle pressure up and down, 111
    operations, 132
    project manager, 120
    QA team, 119, 120
    risk/rewards, 124, 125
    software to data as a product,
        switching, 116, 117
    sponsorship/partnerships, 109
    SQL/ETL developers, 131
    top-down push, 110
    uncertainty, 128
        data sources, 128
        output model, 129
        timestamps/backtracking, 129
    use cases, 126
Business value, 221

C

C-level support, 240
Client algorithms, 233
Colson, Eric, 229
Consulting companies, 181, 219
Continuous integration/continuous
    delivery (CI/CD), 44
Customer service algorithms, 233

D, E

Data analysts, 30
Database administrators
    (DBAs), 9, 36, 49
Data democratization or self-service, 157
Data engineering functions, 14, 44,
    190, 246, 249
Data engineering team, 13, 14, 214
    analysis, 47
    architects, 62, 63
    big data, 55
    CI/CD, 44
    client/server systems, 44

data engineer, 44
data governance, 51
data lineage, 52
data pipelines, 43
  vs. data science team, 58, 59
discovery systems, 52
distributed systems, 46
domain knowledge, 51
expertise level
    new data engineers, 53
    qualified data engineer, 53
    veteran, 53
metadata, 52
misconception
    data scientist, 57, 58
    data warehousing, 56
    data wrangling, 58
placement, 63, 64
programming, 46
retraining existing staff
    ability gap, 59
    nondata engineer, 60
    software engineers, 60
    SQL-focused positions, 61
    technical stack, 59
schema, 50
software engineering, 44
specialization, 54, 55
SQL, 49
verbal communication, 48
visual communication, 47, 48
Data engineers, 220
Data governance, 51
Data hoarding, 105
DataOps team
  cross-functional, 83
  data engineering, 84, 88, 89
  data products, 83

staff, 87
structure, 90
tools, 84
trade-offs, 84, 85
values, 87
Data pipeline, 43
  data flow, 225
  issues in production, 226
  production outages, 224
Data platform team, 158
Data products, 190
  decision-making, 135
  one-off/ad hoc insight, 136
  technologies, 137
Data projects, 214
  issues, 215, 216
  ROI, 215
Data science teams, 12, 13, 30, 159
  client algorithms, 233
  communications, 33
  customer service algorithms, 233
  data scientists
    recruiting, 38
    retraining, 36, 37
    upgrading skills, 37, 38
  distributed systems, 32
  domain knowledge, 33, 34
  goal, 29
  manage inventory, 233
  math, 31
  meeting
    borrowing, 39
    software engineering
      practices, 40
    stifles progress, 40, 41
  operations algorithms, 233
  programming, 32
  skills, 30

Data science teams (*cont.*)
  styling algorithms, 233
  technical debt
    definition, 34
    discovery phase, 34
    Google, 35
    novice approach, 34
    oneitis, 34
    terrible hack/workarounds, 35
Data scientists, 12, 13, 92, 190, 249
  dropping/stopping projects, 213
  operations tasks, 212, 213
Data strategy, 220
Data teams
  beaten path, 106
  challenges, 97
  communicating value, 103
  data scientist, 103
  high-bandwidth connections, 100
  iterative process, 100, 101
  manager/lead engineer, 102
  poor-quality code, 104
  product *vs.* feature teams, 99
  ratios, 98
  self-taught coders, 104
  software engineering, 104
  technical debt, 107
  velocity, 101, 102
Data teams management, 251
Data warehousing, 9
DevOps, 117
Disaster recovery, 73
Distributed systems, 46, 223, 247, 248

F, G

Full-stack people, 27
Funding/resources
  cloud, 123
  data teams, 121
  software/hardware, 121, 122

H

Hadoop computing engines, 245
Heron, 245
Holy grails, 221
Human resources (HR) department
  adding new titles, 169
  C-level, 168
  data engineer, 170
  data scientist, 169
  operation engineer, 170, 171
  titles, 168

I

Interviews
  Bas Geerdink, 253
  British telecommunications
    company (TC)
    background, 269
    business, 270, 275
    data scientist/data engineers, 271
    enterprise infrastructure/
      operations, 273, 274
    initiative, 270
    models, enterprise, 271, 272
    project management
      frameworks, 274
    proof-of-concept, 272, 273
  data engineers, 259, 260
  data science, banking, 258
  Harvinder Atwal,
    Moneysupermarket
    background, 261

barriers and friction,
removing, 263, 264
data teams, changes, 266
hyperparameters, 267, 268
project management
frameworks, 265
team KPIs, 265
team structure, 262
ING data, 254, 255
ING organizational structure, 255–257
KPIs, data teams, 259
Mikio Braun, Zalando
background, 277
business/data teams, 284
data scientists, 283, 284
European and U.S companies, 284
job titles, definitions, 281
KPIs, 283
machine learning, 285
music learning, 279, 280
organizational structure, 278
project management
frameworks, 282
reducing friction, 281, 282
project management framework, 258

J

Java virtual machine (JVM), 145

K

Key Performance Indicators (KPIs)
data engineering team, 133
data science team, 133
operations, 133
technical issues, 130
Klingenberg, Brad, 230

L

Large organizations, 163, 164
Location/status, team
business unit, 155, 156
center of excellence, 156
data democratization/self-service, 157
hub/spoke model, 156
Long-term career, 189
Long-term project
management, 140, 141
Low-cost exploration, 235

M

Machine learning engineers
algorithm, 91
data engineers vs. data scientists, 92
data scientist, 91
high-bandwidth connection, 93
location, 93
Machine learning (ML), 12, 29
Machine-to-machine algorithms, 238
Management task, 27
Massively parallel processing
(MPP), 210
Medium-sized organizations, 162, 163
Metadata, 52
Modern machine learning models, 271

N

Netflix, 241
New teams
critical first hire, 154, 155
data engineers, 152
data scientist, 153
operation engineers, 153
timeline for hirings, 151

Natural language processing (NLP), 169

Noisy neighbor problem, 224

O

Off-the-shelf program, 159

Old guard's management, 188

One-off/ad hoc insight, 136

Operational responsibilities, 234

Operations algorithms, 15, 16, 233

Operations engineer, 16

Operations team
  big data, 65
  cloud *vs.* on-premises
    systems, 81, 82
  cluster software and custom
    software, 65
  data/data quality, 76
  data structures/formats, 71
  disaster recovery, 73
  distributed systems, 66, 69
  framework software, 77
  hardware, 68, 69, 78
  job titles, 67
  monitoring/instrumenting, 72, 73
  organization's code, 76
  scripting/programming, 71
  security, 70
  SLAs (*see* Service-level
    agreements (SLAs))
  software, 69
  troubleshooting, 70, 75
  use cases, 72

Organizational changes, 187
  nondata team, 189
  old guard, 187, 188

Organizational issues, 216

Organizations architectures, 222

P

Personally identifiable
    information (PII), 51

Petabyte-scale data, 232

Platform team, 234

Processes/values, 239

Project management framework,
    236, 237, 250

Q

Quality assurance (QA)
    team, 119, 184, 197

R

Ramasamy, Karthik, 245

Real-time SLAs, 74

Reporting structure
  CEO, 165
  CIO, 166
  CTO, 165
  nonoverlapping/conflicting
    goals, 164
  VP, engineering, 165
    CDO, 167
    data, 167
    finance, 167
    marketing, 166
    product, 166

Ryaboy, Dmitriy, 243
  data and analytics, 245, 246
  data engineer, 249
  data scientist, 249
  data team, 250
  project management, 250
  task ownership, 247
  Twitter, 243

S

Service-level agreements (SLAs)
  batch, 74
  organization code/
    deployment, 75
  real-time, 74
  response time, 73
  specific service/technology, 75
Silver bullets, 214
Site reliability engineer (SRE), 67
Skill/ability gaps
  beginners, 210, 211
  cluster, 211
  differences, 207, 208
  distributed systems, 209
  meaning, 207
  MPP, 210
  programming ability, 208
Skills gap assessment
  ability gaps, 194, 195
  data scientists, 198
  definition, 191
  experience levels, 191
  gap analysis, 193, 194
  hardware requirements, 195
    big teams, 196
    cloud, 195
    development, 196, 197
    purchase clusters, 195
    test/QA, 197
  spreadsheet, 192
Small organizations
  business intelligence team, 160
  consequence, 161
Software engineering, 9, 10, 104
SQL, 49
Stack Overflow, 106
Staffing, operation team

data engineer, 80
DevOps task, 79, 80
nontrivial task, 79
training, 78
Stitch Fix, 229
  algorithms department, 230
  architecture, 234
  capabilities, 232
  clients, 237
  competitive advantage, 237
  controlled experiments, 239
  data scientists, 231
  management structure, 231
  merchandising areas, 231
  organizational structure, 231
  production systems, 234
Streaming Compute, 245
Stuck teams
  big/small data, 204
  project complexity, 203
  root causes, 202
  technical difficulty, 203
Styling algorithms, 233

T

Technology choices, 182
  big data projects, 148
  distributed
    systems, 142, 143
  error bars, 147
  mental framework, 141, 142
  order consequences, 149
  programming languages, 144
    data engineers, 145, 146
    data scientist, 145
  project management
    framework, 146, 147

Technology failure, 216
Traditional database technologies, 245
Training, 180

U

Underperforming teams
    complex task, 206, 207
    development level, 205
    meaning, 204

straightforward task, 206
vs. stuck teams, 204
unusable data, 206

V, W, X, Y, Z

Velocity, 101
Vendor, 219
Verbal communication, 48
Visual communication, 47, 48

Printed in the United States
by Baker & Taylor Publisher Services